Technology of Welding and Joining

Technology of Welding and Joining

Editor

Tomasz Wegrzyn

MDPI • Basel • Beijing • Wuhan • Barcelona • Belgrade • Manchester • Tokyo • Cluj • Tianjin

Editor
Tomasz Wegrzyn
Faculty of Transport and
Aviation Engineering,
Silesian University of Technology
Poland

Editorial Office
MDPI
St. Alban-Anlage 66
4052 Basel, Switzerland

This is a reprint of articles from the Special Issue published online in the open access journal *Metals* (ISSN 2075-4701) (available at: https://www.mdpi.com/journal/metals/special_issues/welding_joining).

For citation purposes, cite each article independently as indicated on the article page online and as indicated below:

LastName, A.A.; LastName, B.B.; LastName, C.C. Article Title. *Journal Name* **Year**, *Volume Number*, Page Range.

ISBN 978-3-0365-0886-3 (Hbk)
ISBN 978-3-0365-0887-0 (PDF)

© 2021 by the authors. Articles in this book are Open Access and distributed under the Creative Commons Attribution (CC BY) license, which allows users to download, copy and build upon published articles, as long as the author and publisher are properly credited, which ensures maximum dissemination and a wider impact of our publications.

The book as a whole is distributed by MDPI under the terms and conditions of the Creative Commons license CC BY-NC-ND.

Contents

About the Editor vii

Preface to "Technology of Welding and Joining" ix

Tomasz Węgrzyn, Tadeusz Szymczak, Bożena Szczucka-Lasota and Bogusław Łazarz
MAG Welding Process with Micro-Jet Cooling as the Effective Method for Manufacturing Joints for S700MC Steel
Reprinted from: *Metals* 2021, 11, 276, doi:10.3390/met11020276 1

Zhanzhan Su, Zhengqiang Zhu, Yifu Zhang, Hua Zhang and Qiankun Xiao
Recrystallization Behavior of a Pure Cu Connection Interface with Ultrasonic Welding
Reprinted from: *Metals* 2021, 11, 61, doi:10.3390/met11010061 19

Iñigo Calderon-Uriszar-Aldaca, Estibaliz Briz, Harkaitz Garcia and Amaia Matanza
The Weldability of Duplex Stainless-Steel in Structural Components to Withstand Corrosive Marine Environments
Reprinted from: *Metals* 2020, 10, 1475, doi:10.3390/met10111475 37

Jiankun Xiong, Ting Li, Xinjian Yuan, Guijun Mao, Jianping Yang, Lin Yang and Jian Xu
Improvement in Weldment of Dissimilar 9% CR Heat-Resistant Steels by Post-Weld Heat Treatment
Reprinted from: *Metals* 2020, 10, 1321, doi:10.3390/met10101321 61

Chuanguang Luo, Huan Li, Yonglun Song, Lijun Yang and Yuanhua Wen
Microstructure and Performance Analysis of Welded Joint of Spray-Deposited 2195 Al-Cu-Li Alloy Using GTAW
Reprinted from: *Metals* 2020, 10, 1236, doi:10.3390/met10091236 71

Guoqian Mu, Wenqing Qu, Haiyun Zhu, Hongshou Zhuang and Yanhua Zhang
Low Temperature Cu/Ga Solid–Liquid Inter-Diffusion Bonding Used for Interfacial Heat Transfer in High-Power Devices
Reprinted from: *Metals* 2020, 10, 1223, doi:10.3390/met10091223 83

Bokai Liao, Hong Wang, Shan Wan, Weiping Xiao and Xingpeng Guo
Electrochemical Migration Inhibition of Tin by Disodium Hydrogen Phosphate in Water Drop Test
Reprinted from: *Metals* 2020, 10, 942, doi:10.3390/met10070942 97

Zhenjiang Wang, Zeng Gao, Junlong Chu, Dechao Qiu and Jitai Niu
Low Temperature Sealing Process and Properties of Kovar Alloy to DM305 Electronic Glass
Reprinted from: *Metals* 2020, 10, 941, doi:10.3390/met10070941 109

Ashutosh Sharma, Min Chul Oh, Myoung Jin Chae, Hyungtak Seo and Byungmin Ahn
Solderability, Microstructure, and Thermal Characteristics of Sn-0.7Cu Alloy Processed by High-Energy Ball Milling
Reprinted from: *Metals* 2020, 10, 370, doi:10.3390/met10030370 123

Aleksander Lisiecki and Dawid Ślizak
Hybrid Laser Deposition of Fe-Based Metallic Powder under Cryogenic Conditions
Reprinted from: *Metals* 2020, 10, 190, doi:10.3390/met10020190 135

Jian Qiao, Peng Yu, Yanxiong Wu, Taixi Chen, Yixin Du and Jingwei Yang
A Compact Review of Laser Welding Technologies for Amorphous Alloys
Reprinted from: *Metals* **2020**, *10*, 1690, doi:10.3390/met10121690 . . . . **155**

Qi Zhu, Miaoxia Xie, Xiangtao Shang, Geng An, Jun Sun, Na Wang, Sha Xi, Chunyang Bu and Juping Zhang
Research Status and Progress of Welding Technologies for Molybdenum and Molybdenum Alloys
Reprinted from: *Metals* **2020**, *10*, 279, doi:10.3390/met10020279 . . . . **169**

About the Editor

Tomasz Wegrzyn is a welding specialist. He defended his doctorate in 1991 at the Silesian University of Technology, did his habilitation in 2001 at the Warsaw University of Technology, and was nominated by the President of Poland as a full professor in 2016. He researches the welding of various grades of steel and nonferrous alloys and is the author of 250 publications. He has promoted 8 doctors of Science. He has worked in 3 European scientifical centers: Institute Polytechnic in Guarda (Portugal), University Da Beira Interior (Covilha, Portugal), and Silesian University of Technology (Katowice, Poland). He lectures in the field of welding and materials engineering and works in the field of Civil Engineering and Transport. His greatest achievements include works in welding metallurgy and welding technology. He studied the influence of alloying elements and welding conditions on the nucleation of acicular ferrite in low alloy steels. He has also analysed welding problems of grades of steel. For the first time, he proposed the classification of welding methods into various oxygen and various nitrogen processes. He is the main co-creator of a new welding process with microjet cooling. He is the author of many patents and technological solutions.

Preface to "Technology of Welding and Joining"

In this book, you will find information on new materials and new welding technologies. Problems related to the welding of difficult-to-weld materials are considered and solved. The latest welding technologies and processes are presented. This book provides an opportunity to learn about the latest trends and developments in the welding industry. Enjoy reading.

Tomasz Wegrzyn
Editor

MAG Welding Process with Micro-Jet Cooling as the Effective Method for Manufacturing Joints for S700MC Steel

Tomasz Węgrzyn [1,*], Tadeusz Szymczak [2], Bożena Szczucka-Lasota [1] and Bogusław Łazarz [1]

1. Faculty of Transport and Aviation Engineering, Silesian University of Technology, 40-119 Katowice, Poland; Bozena.Szczucka-Lasota@polsl.pl (B.S.-L.); Boguslaw.Lazarz@polsl.pl (B.Ł.)
2. Motor Transport Institute, ITS Jagiellońska 80, 03-301 Warszawa, Poland; tadeusz.szymczak@its.waw.pl
* Correspondence: tomasz.wegrzyn@polsl.pl; Tel.: +48-504-816-362

Abstract: Advanced high-strength steel (AHSS) steels are relatively not very well weldable because of the dominant martensitic structure with coarse ferrite and bainite. The utmost difficulty in welding these steels is their tendency to crack both in the heat affected zone (HAZ) and in weld. The significant disadvantage is that the strength of the welded joint is much lower in comparison to base material. Adopting the new technology regarding micro-jet cooling (MJC) after welding with micro-jet cooling could be the way to steer the microstructure of weld metal deposit. Welding with micro-jet cooling might be treated as a very promising welding S700MC steel process. Tensile and fatigue tests were mainly carried out as the main destructive experiments for examining the weld. Also bending probes, metallographic structure analysis, and some non-destructive measurements were performed. The welds were created using innovative technology by MAG welding with micro-jet cooling. The paper aims to verify the fatigue and tensile properties of the thin-walled S700MC steel structure after welding with various parameters of micro-cooling. For the first time, micro-jet cooling was used to weld S700MC steel in order to check the proper mechanical properties of the joint. The main results are processed in the form of the Wöhler's S–N curves (alternating stress versus number cycles to failure).

Keywords: civil engineering; mechanical engineering; transport; smart city; micro-jet welding; mechanical resistance; microstructure; mini-specimen; fracture; fatigue limit

1. Introduction

A smart city is an urban area that uses different types of electronic methods and sensors to collect data [1]. This includes also data about transportation systems and communication technology [2–4]. Transport based on lighter and more durable means of transport will play an important role in the smart city. Authors put especial attention to the necessity of using modern materials and the development of various technologies (including welding) to create a Smart City model. In the modern automotive industry, high-strength martensitic steels from the AHSS group are increasingly used. Still welding of AHSS steels does not meet the satisfactory results because of the much lower tensile [5,6] and fatigue strength [6–8] of the joint compared to the base material. The differences of these properties are the consequence of chemical composition and structure of the base material and weld, which results in a need to use electrode wires with an increased nickel and molybdenum content and much lower sulfur content that affects the joint's structure and mechanical properties. The weld metal deposit contains mainly martensite and bainite with coarse ferrite while the base material contains mainly martensite and rather much fragmented ferrite. S700MC advanced high-strength steel (AHSS) can be obtained as hot-rolled, cold-rolled, hot-dip galvanized, and electro galvanized products. The material range includes thin sheet steel with a thickness range from 0.4 mm to 3 mm. This form of steel is used in slimming various structures (mobile platforms, containers, etc.,).

The increase in the use of modern AHSS steels in the automotive industry results from the possibility of reducing the thickness of car body sheets with a simultaneous improvement in the mechanical properties of the structure compared to the use of conventional low alloy steels. For construction of various modes of transport, the usage of high-strength martensitic steels from the group becomes increasingly popular [9–12]. These steel grades are relatively not very well weldable. That joint has good mechanical properties after welding, especially high impact toughness and elongation. The weld is not prone to cracks. The utmost difficulty in welding these steels is their tendency to crack in the weld and heat affected zone [8–10]. The main parameter determining the weldability of the steels is the carbon content and the carbon equivalent. In S700MC steels, the CET coefficient (carbon content equivalent, which, apart from C, also takes into account the impact on the weldability of steel of other elements: Mn, Mo, Cr, Cu, Ni) is similar to that of low-alloy steels [13].

This study analyzes the welding possibilities of AHSS steel with micro-jet cooling for the first time [7,14,15]. Welding high-strength steels is difficult. It is important to properly select the process and determine the correct parameters. The experimental results showed that the laser HSLA welds failed in a ductile necking/shear failure mode and the ductile failure was initiated at a distance away from the crack tip near the boundary of the base metal and heat affected zone [16]. Micro-jet cooling provides different cooling conditions than classic "macro" cooling. It is related to the so-called "scale effect," mentioned by other authors [17,18], who found, for example a great influence of micro-jet cooling on ferrite grinding during welding of low-alloy steel, which could not be obtained during "macro" cooling. After micro-jet cooling in the same welding conditions 70% of acicular ferrite was obtained, while after welding with classic "macro" cooling it was possible to obtain only 50% acicular ferrite, the most favorable phase guaranteeing high impact toughness at low temperatures and attractive ductility. For this reason, there was no attempt to deal with macro cooling in this research. The micro-jet cooling enables for selective and spot cooling of the welds, which allows controlling the microstructure of the welds. This method has proven successful in welding low-alloy steels and aluminum alloys [19].

Because of the welding with micro-jet cooling in welded joints made of low-alloy steels, high impact strength at low temperatures is obtained. In the case of aluminum alloy, higher strength and much better electrical conductivity (what was the purpose of this research) is reached as the effect of the joining process [18].

Fatigue and tensile tests are important destructive experiments in the quality assessment of AHSS welds [20–23]. Generally, during AHSS steels welding, a thermodynamic analysis of the process is required. In order to reduce welding stresses and to refine the ferrite, it is recommended to limit the linear energy during welding up to 4 kJ/cm [24–26]. In order to reduce the hydrogen content in the weld, it is recommended to use preheating and control the temperature of the interpass layers [27,28]. During welding, single hydrogen atoms H are formed in the weld, where they combine with the H_2 molecule (recombination effect). Accumulation of hydrogen inside the metal causes the formation of internal pressure, causing internal stresses of the material, which in turn lead to the formation of hydrogen-induced cracking (HIC) [29–34]. HIC cracking in S700MC steel is initiated mainly at the martensite-ferrite grain boundaries and in contact with ferrite [35–38]. In order to improve the process of increasing its repeatability, it was decided to weld S700MC steel with various parameters of micro-jet cooling. Micro-jet cooling consists in passing a narrow stream of gas through a specially designed injector, coupled to the welding head. In welding conditions, the following gases are used for this process: helium, argon, carbon dioxide, air, nitrogen, various mixtures. The gas pressure is in the range 0.4–0.7 MPa and the stream diameter is in the range of 50–80 μm.

Steel S700MC steel joints were not made using micro-jet cooling, so it is important to know the fatigue strength of this steel, because fatigue strength is one of the fundamental data for concluding on the quality of the joint [39–43]. This is an important reason for the research undertaken.

2. Materials and Methods

Welded (BW) butt joints of S700 MC steel with a thickness of 3 mm were made. The MAG (135) welding method in the flat position (PA) was applied according to the requirements of EN 15614-1 standard. Preparation of the material for a single-stitch welding and a finished 3-mm thick weld is shown in the Figure 1.

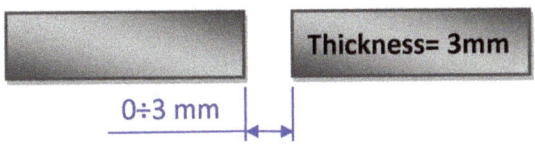

Figure 1. Preparation of the element for metal active gas (MAG) welding with micro-jet cooling.

It was decided to produce welds with the use of MAG (metal active gas) process testing the following gas mixture acting as a shielding gas mixture: Ar + 18% CO_2 (according to the PN-EN 14175 standard). Two different electrode wires (recommended for welding S700MC steel) were used for welding in order to more accurately investigate the effect of micro-jet cooling on the mechanical properties of a joint made in various welding conditions. The use of two different wires helps to judge that micro-jet cooling is beneficial when welding this steel. All of the samples were welded with two electrode wires (UNION X90 and UNION X96):

- EN ISO 16834-A G 89 6 M21 Mn4Ni2CrMo-UNION X90 (C 0.10, Si 0.8, Mn 1.8, Cr 0.35, Mo 0.6, Ni 2.3);
- EN ISO 16834-A G 89 5 M21 Mn4Ni2,5CrMo-UNION X96 (C 0.11, Si 0.78, Mn 1.9, P 0.01, S 0.009, Cr 0.35, Mo 0.57, Ni 2.23, V 0.004, Cu 0.02, Ti 0.057, Zr 0.001, Al 0.002).

Welded joints were made by MAG welding (135) using the Migatronic MIG-A Twist burner (Migatronic, Fjerritslev, Denmark) with intelligent arc control function. The burner is equipped with the IAC™ Intelligent Arc Control function, which provides a lower value of input energy, less distortion, while maintaining the mechanical properties of the material. The used technology ensures much lower power consumption compared to traditional welding machines [44]. Welding tests were performed in the gas shielding mixture M21 (82% Ar + 18% CO_2) with the assumed gas flow of 13 l/min. The input energy during the welding of thicker sheets (3 mm) was below the recommended value of 4 kJ/cm. All welding tests were carried out without preheating (Table 1). Chemical composition and mechanical properties of the S700MC steel are presented in Table 1 [45]. The gap between sheets was varied in the range 0–3 mm. The best results were obtained for gap = 1.5 mm. This case was taken for further tests only.

Table 1. Welding parameters of S700MC steel.

Layers Order	Welding Method	Diameter of the Electrode, mm	Current Intensity, A	Voltage, V	Polarization	Welding Speed, m/min	Input Energy, kJ/cm
1	135	1.0	100	19	DC "+"	300	below 4

Micro-jet cooling parameters were slowly varied:

- Micro-jet gas: argon
- Stream diameter: 60 and 70 μm
- Gas pressure: 0.6 and 0.7 MPa.

Once all of the test were performed the following quality control checks were applied: non-destructive and destructive tests.

Non-destructive tests included (NDT):

- Visual testing (VT) of the manufactured welded joints was performed with an eye armed with a loupe (Levenhook, Tampa, FL, USA) at 3× magnification—tests were carried out in accordance with the requirements of the PN-EN ISO 17638 standard, evaluation criteria according to the EN ISO 5817 standard.

Visual testing of welds was made using standard auxiliary measures, luxmeter with white light 520 Lx. It was found that only some of the tested welds were made correctly and met the quality requirements; they were characterized by the limit of acceptability "B" according to PN-EN ISO 5817 [43]. Magnetic-particle test of welds was made using the wet method with the following conditions: field strength 3 kA/m, white light 515 Lx, temperature 20 °C, MR-76 detection means, MR-72 contrast.

- Magnetic-particle testing (MT)—the tests were carried out in accordance with the PN-EN ISO 17638 standard, the evaluation of the tests was carried out in accordance with the EN ISO 5817 standard, the device for testing was a magnetic flaw detector of REM—230 type (ATG, Prague, Czech Republic).

The destructive tests included:

- Visual tests on micro-sections of welded joints were performed with an eye armed with a loupe at 3× magnification—tests were performed according to PN-EN ISO 17638 standard with reagents for testing according to PN-CR 12361 standard, evaluation criteria according to EN ISO 5817 standard;
- The bending test was carried out in accordance with the PN-EN ISO 5173 standard, using the ZD-40 testing machine (WPM, Leipzig, Germany).

Then, a bending test was performed only for those specimens that passed the NDT test. Thus bending test was carried out only for specimens without cracks (with B acceptability). The tests used: specimens with a thickness of a = 3 mm, width b = 20 mm, mandrel d = 22 mm, spacing of supports d + 3a = 31 mm, and the required angle of bending 180. Five bending test measurements were carried out for each tested joint thickness on the root side and on the face side. Specimens welded with both electrode wires (UNION X90 and UNION X96) with two variants (with micro-jet cooling and without micro-jet cooling) were selected for the tests: with (a) argon micro-jet cooling with stream diameter 60 µm and gas pressure 0.7 MPa, wire UNION X90, (b) with argon micro-jet cooling with stream diameter 60 µm and gas pressure 0.7 MPa, wire UNION X96, (c) without micro-jet cooling, wire UNION X90, (d) without micro-jet cooling, wire UNION X96.

- Examination of microstructure of specimens etched with Adler reagent using light microscopy (Neophot 32, Carl Zeiss Jena, Jena, Germany);

Samples prepared for observation after welding with electrode wire UNION X90 with two variants (with micro-jet cooling and without micro-jet cooling). The area of the parameters of the preferred micro-jet cooling was narrowed down again for the analysis of the M90-60-07 sample and the W90 sample that corresponded with: (a) MAG welding with argon micro-jet cooling with stream diameter 60 µm and gas pressure 0.7 MPa, wire UNION X90, (b) MAG welding without micro-jet cooling, wire UNION X90.

- Tensile and fatigue tests.

Tensile and fatigue tests were conducted using the 8874 INSTRON servo-hydraulic testing machine (Instron, High Wycombe, UK) and mini-specimens at room temperature. The following standards PN-EN ISO 6892-1:2020 [46] and ASTM E468-18 [47] were used, respectively. The specimens were directly selected from parent material and region with weld, manufacturing flat (Figure 2a) and hourglass specimens (Figure 2b) for monotonic and cyclic tension tests, respectively. Tensile experiments were carried out by means of three types of loading signal i.e., displacement, strain, and stress, which have the following values of velocity: 1 mm/min, 0.08 1/min, and 178 MPa/min, respectively. The axial strain of specimens subjected to tension was captured by means of an axial extensometer on gauge length of 12.5 mm (Figure 2a). The specimens for fatigue tests were designed basing on the requirements of the following standards: ASTM E466-15 [48] and ASTM E468-18 [47]. The face and root of the mini-specimens were removed in technological process with respect to examining the weld at a defined stress state. Worth to notice that results on the specimens with a smooth measuring section can be easy compared with the data on parent material. Moreover, the removal of the face and root of the weld allows the quality of the joint to be assessed directly in the zone of the used parts. Nevertheless, from the engineering point of view, both kinds of specimens should be used because results of modelling and technical data as well as elaborating stages of inspections can be covered easier. In the proposed experimental procedure only specimens with the smooth measurement section were employed with respect to this feature which appears in a lot of specimens used in experimental mechanics of materials and its degradation can be followed at a small number of technical limits. Moreover, the measuring region in the form of an hourglass has directly enabled testing the weld under selected loading conditions. The weld zone was of the physical fracture plane for damages initiation, coalescence, and growing up to fracture. This kind of specimen is also much less sensitive to any scratches and big roughness because of the narrow fracture region. Therefore, this approach was used to shape and prepare the specimens. Worth to emphasis that from a practical point of view, experiments on welded specimens having a face and root can follow the additional piece of the knowledge because all sections of a joint can be tested. Comparing data from experiments on machined [49] and non-machined specimens [50], differences in weld behavior can be evidenced [51,52]. Nevertheless, this stage will be considered in the next paper, as an additional approach to examining the micro-jet cooling welded joints of S700MC steel.

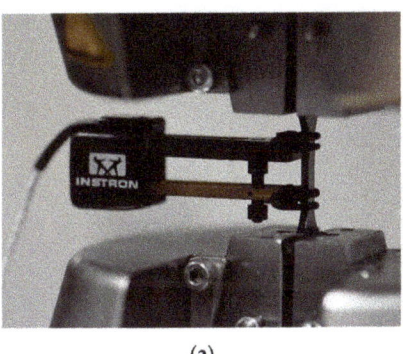
(a)

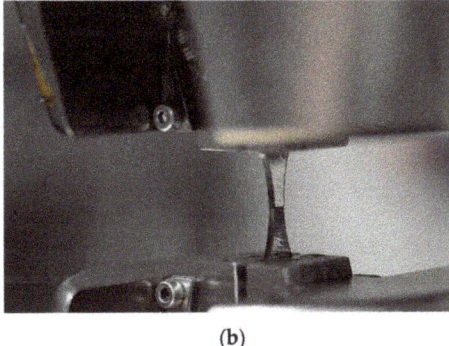
(b)

Figure 2. Flat (**a**) and hourglass (**b**) mini-specimen in testing machine before tensile and fatigue test, respectively, nominal dimensions of the cross sections in measuring zones: 4 mm × 2 mm (**a**); 4 mm × 3 mm (**b**).

In the case of fatigue test, with respect to dynamic fracturing of the weld tested, which would lead to failure of the extensometer, a displacement collected by the sensor of the testing machine was used for assessing the weld deformation. All tests were performed employing stress signal at a maximum value between of 400 MPa and 950 MPa at stress ratio $R = \sigma_{min}/\sigma_{max} = 0.1$ and frequency 10 Hz. During fatigue examination, values of the following physical quantities were recorded: force (Figure 3(a1)), stress (Figure 3(b1)), displacement (Figure 3(a2,b2)), total energy (Figure 4), time, and cycles. Data in a form of displacement versus a number of cycles were taken to assess the weld behavior under cyclic loading, expressed by cyclic hardening and softening at initial and subsequent cycles, respectively (Figure 3(a2,b2)). They were collected in various types of files, which follows: courses of the physical quantities, their maximal and minimal values, and values directly before and during fracture. In the case of the last stage, the 50th last cycles before decohesion were used for determining the weld behavior at the final stage of the material tested. With respect to extending the knowledge on the weld behavior under stress levels, fracture regions were analyzed using the macro-photography technique.

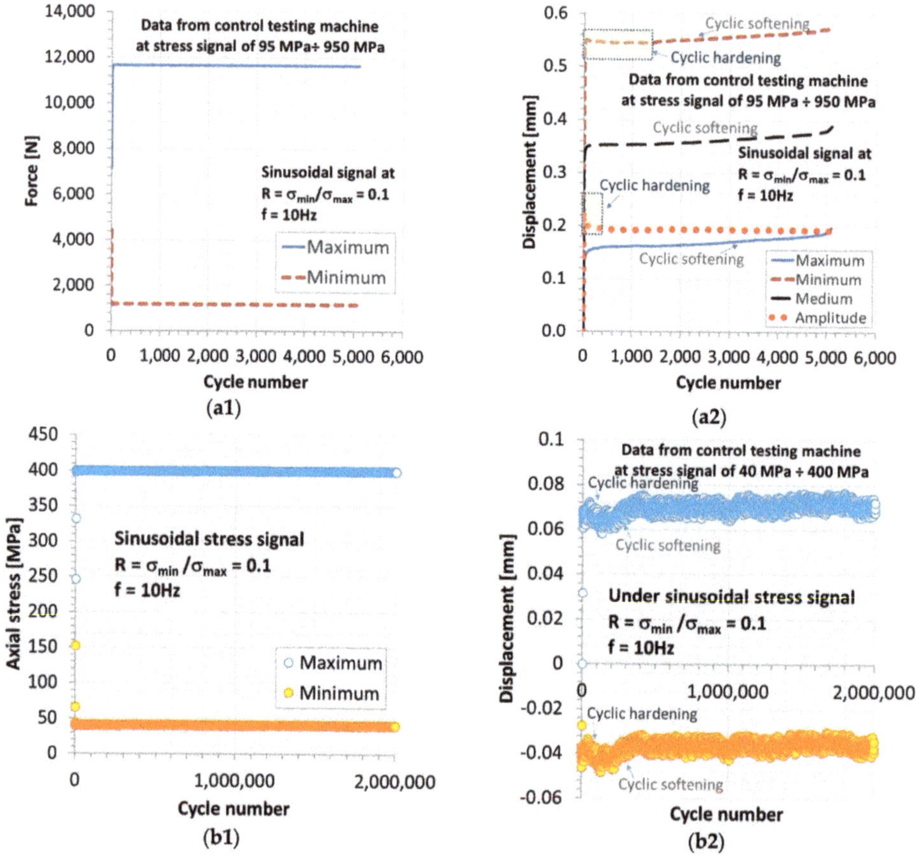

Figure 3. Maximum and minimum values of force (**a1**) and stress signal (**b1**) and displacement (**a2,b2**) from the fatigue test for determining fatigue limit of S700MC steel welded by means of the welding wire of Union X96, reaching 5×10^3 (**a1,a2**) and 2×10^6 cycles (**b1,b2**).

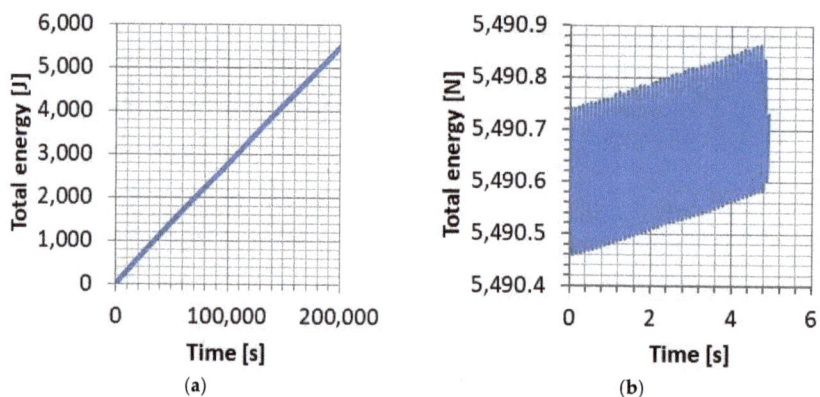

Figure 4. Courses of total energy collected in fatigue test up to fatigue limit (at 2×10^6 cycles) of S700MC steel welded using the Union X96 wire: (**a**) from the beginning of the test up to the stage directly before of the limited number of cycles, (**b**) the last 5 seconds at the final stage of the test up to the moment determining fatigue limit i.e., 400 MPa.

3. Results

3.1. The Results of Non-Destructive Tests

Non-destructive test results are presented in Table 2. They expressed differences in the weld quality and confirmed a meaning of this stage concerning assessment of the quality of zone examined.

Table 2. Assessment of non-destructive testing of the movable platform joint.

Sample Designation	Micro-Jet Stream Pressure MPa	Micro-Jet Stream Diameter μm	Wire	Micro-jet Gas	Observation, Acceptability
W90	without	without	Union 90	without	Cracks in the weld
M90-60-06	60	0.6	Union 90	Ar	Cracks in the weld
M90-60-07	60	0.7	Union 90	Ar	No cracks, B
M90-70-06	70	0.6	Union 90	Ar	No cracks, B
M90-70-07	70	0.7	Union 90	Ar	Cracks in the weld
W96	without	without	Union 96	without	Cracks in the weld
M96-60-06	60	0.6	Union 96	Ar	Cracks in the weld
M96-60-07	60	0.7	Union 96	Ar	No cracks, B
M96-70-06	70	0.6	Union 96	Ar	No cracks, B
M96-60-06	70	0.7	Union 96	Ar	Cracks in the weld

3.2. The Bending Tests

The test results are summarized in Table 3. From the analysis of the results presented in Table 3, it follows that the test was carried out correctly, the evaluation of the tests is positive only in some cases when micro-jet cooling was not too intense and not too weak. Process was correct only with medium power micro-jet cooling, because no cracks and other disconformities were found in the samples tested. By changing the welding parameters (current, welding speed), it is possible to avoid cracks, but it is not possible to obtain high plastic properties of welds with the relative elongation at the level of base material properties. This was examined in microstructural analysis for the weld tested as well as fracture zones after fatigue tests by means of micro-photographic techniques. These approaches did not express any cracks in the weld, which is indicated by the pure quality of the region manufactured. It reflected the high-quality level of the joint and it was confirmed in comparison to the regimes of the PN-EN ISO 5817 standard [52] on a crack length ≥ 0.5 mm at the highest requirement denoted by letter B.

Table 3. Bending tests results.

Sample Designation	Deformed Side	a_0 (mm) × b_0 (mm)	Bending Angle (°)	Notes
M90-60-07	root of weld	3.0 × 20.0	180	no cracks
M90-60-07	face of weld	3.0 × 20.0	180	no cracks
M96-60-07	root of weld	3.0 × 20.0	180	no cracks
M96-60-07	face of weld	3.0 × 20.0	180	no cracks
W90	root of weld	3.0 × 20.0	180	cracks
W90	face of weld	3.0 × 20.0	180	cracks
W96	root of weld	3.0 × 20.0	180	cracks
W96	face of weld	3.0 × 20.0	180	cracks

3.3. Metallographic Examination

Observations of the samples digested in Adler's reagent were carried out on the Reichert light microscope. The examined joints are dominated by a martensitic and ferritic structure—Figure 5 shows the structure of the W90 sample.

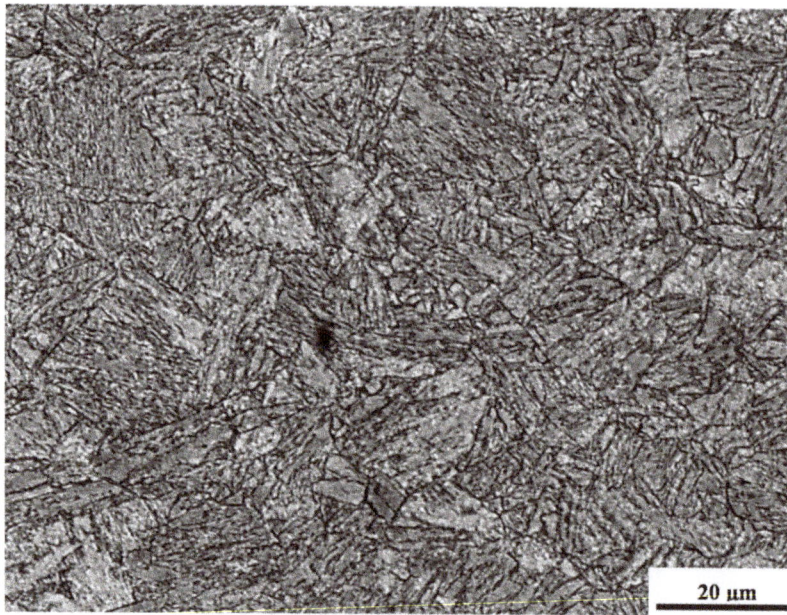

Figure 5. Microstructure of the joint (W90). Visible martensite and course ferrite (LM).

In addition to martensite, coarse-grained ferrite was observed. Only the application of micro-jet cooling after MAG welding made it possible to obtain much more fragmented ferrite (shown in Figure 6).

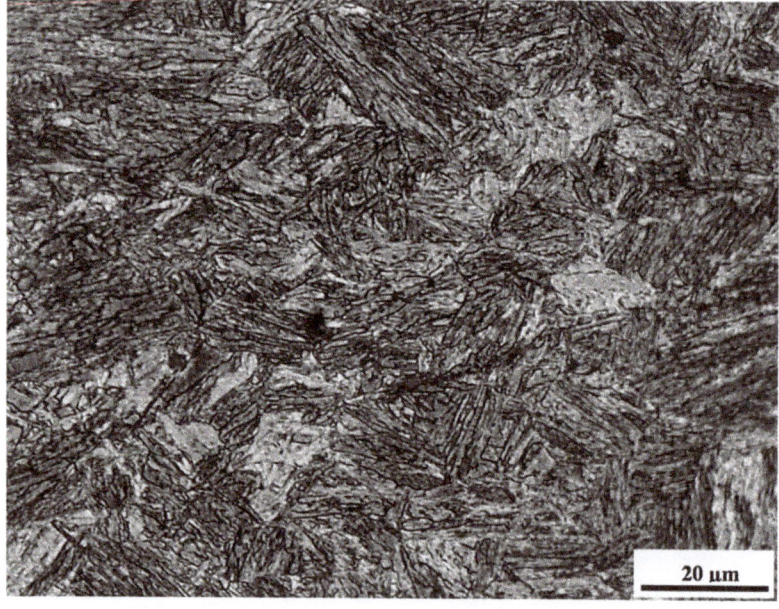

Figure 6. Microstructure of the joint (M90-60-07) after welding with micro-jet cooling. Visible martensite and fine-grained ferrite (LM).

Looking at Figures 5 and 6, it can be concluded that the structure appears to be similar. In Figure 6, it is visible that the ferrite is slightly more fragmented. This is due to the use of micro-jet cooling during welding, which affects the grain growth during the austenitic transformation. Other authors also draw attention to the possibility of grain grinding during welding of steel with the use of micro-jet cooling [17,19].

An important role of microstructure in the steel behavior under loading can be evidenced in the tensile characteristic, which expressed values of yield stress and ultimate tensile strength above 800 MPa at attractive ductility taking of 17.5%, indicating beneficial features of the steel tested in comparison to other types of this kind of material (Figure 7a). This kind of microstructure also results in the stability of the stress parameters (Figure 7b) and similarity of fracture zones (Figure 8) at various types of loading signal, containing: displacement, strain, and stress. It means that in the case of the weld, this type of microstructure is very desired and its manufacturing extends the application range of components with this kind of welded regions, containing different types of loading.

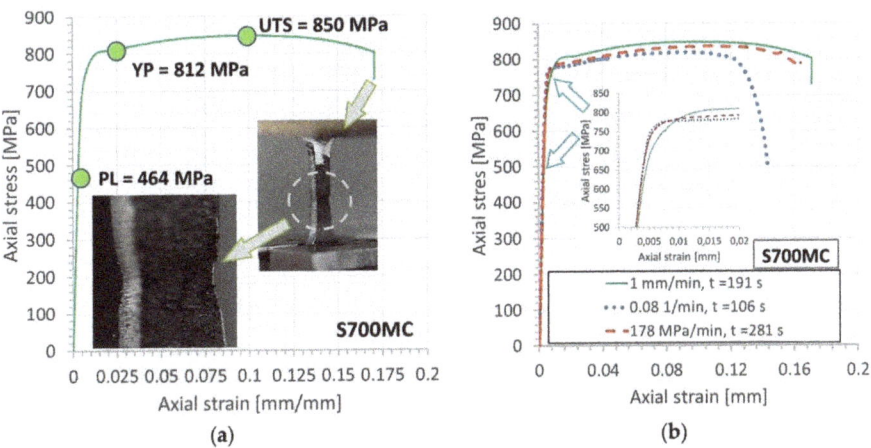

Figure 7. Tensile characteristics of S700MC steel from tests conducted at various loading signals at the following values of velocity for: (**a**) displacement 1 mm/min, (**b**) strain—0.08 1/min.

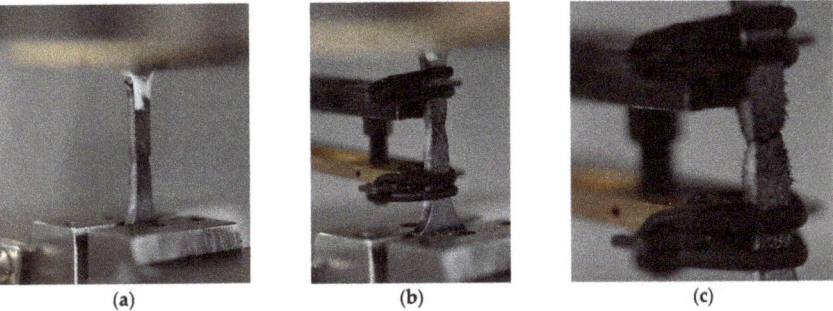

Figure 8. S700MC steel at final stage of tensile tests conducted at the following values of velocity of loading signal: (**a**) 1 mm/min (displacement), (**b**) 0.08—1/min (strain), (**c**) 178 MPa/min (stress).

As it was determined in the tensile tests conducted on the S700MC steel joint produced by the use of the welding wire named the Union NiMoCr, EDFK 1000 and Union X96, that all results exhibited very similar values of ultimate tensile strength and they were only 5% lower than the value of this parameter for the parent material (Figure 9a). Differences in elasto-plastic region up to the necking point are more than evident. Nevertheless, fracture zones expressed variations in the mechanism of degradation from shear fracture (at the welding wires of Union NiMoCr and EDFK 1000) to ductile fracture (for the Union 9 welding wire) (Figure 10a,b, respectively). In comparison with the features of the fracture region of the parent material after tension (Figure 8) the ductile degradation is dominant in the region examined. A significant improvement in plastic properties was demonstrated in relation to the achievements of previous researchers. The relative elongation of the joint was achieved at the level of the parent material, which other authors did not succeed. Therefore, the weld has been selected for tests under cyclic loading (Figures 2b, 3, 4, 11–14). Nobody has analyzed the welding of S700MC steel using an innovative process using micro-jet cooling so far, and nobody has checked the fatigue strength of such welds. Results captured from this kind of tests are represented by fracture regions (Figure 11), variations of displacement (Figure 12), and total energy (Figure 13) at final cycles as well as the Wöhler curve (Figure 14).

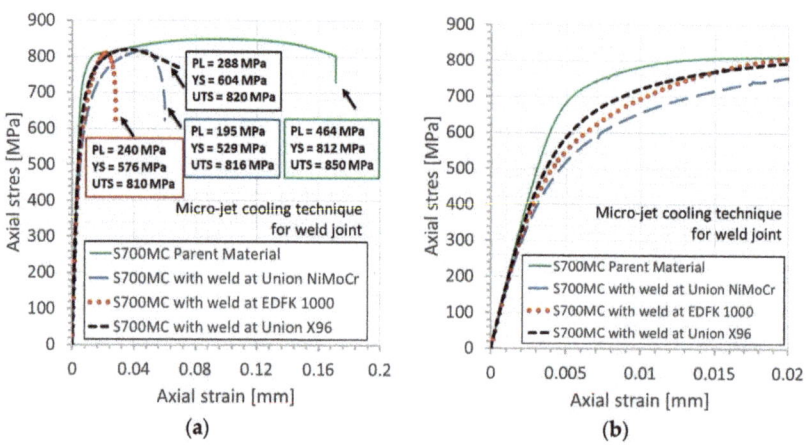

Figure 9. Tensile characteristics of S700MC welded by means of three different welding wires in: (a) general view, (b) focuses on elastic and elastic-plastic regions.

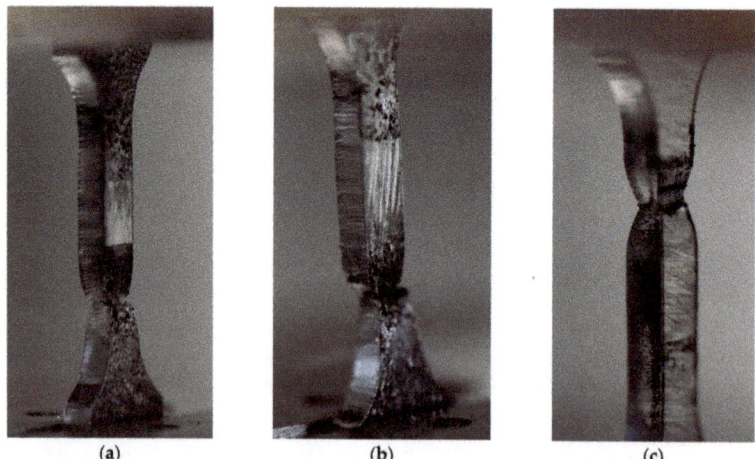

Figure 10. Fracture zones of tensile specimens with joints manufactured at the welding wires: (a) Union NiMoCr, (b) EDFK 1000, (c) Union X96.

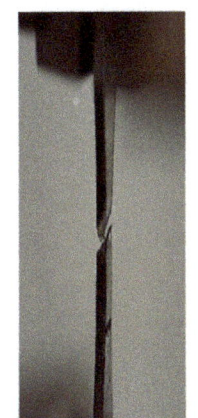 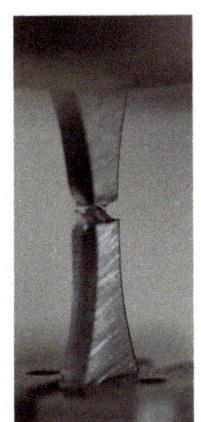 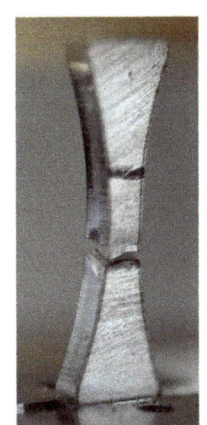 

(a) Maximum stress = 950 MPa (b) Maximum stress = 800 MPa (c) Maximum stress = 700 MPa (d) Maximum stress = 600 MPa

Figure 11. Fracture zones of S700MC steel welded at supporting the micro-jet cooling technique at various values of maximal stress after fatigue test, i.e. (**a**) 950 MPa, (**b**) 800 MPa, (**c**) 700 MPa and (**d**) 600 MPa.

Data captured from fatigue tests have enabled to follow difference in the degradation of the weld examined at various stress levels (Figure 11). They were represented by reorientation of fracture zone from angular (Figure 11a–c) to horizontal (Figure 11d) at a high and low value of stress applied, respectively. Connecting these results with tensile mechanical properties of the weld (Figure 9) the stress value related to the change of degradation mechanism can be indicated. In this case, the yield stress plays this role being a major factor in the weld behavior under the stress range considered.

Some differences in the weld response on the stress values applied are collected based on data in a form of displacement versus time (Figure 12), but they were not significant as taken from the fracture regions. They show a limit value of displacement before the appearance of weld fracture, i.e., 3 mm, indicating on the almost same level of deformation at crack growing. This occurs under unloading, resulting in a rapid reveal of the fracture zone. Nevertheless, the final stage connected with decohesion appeared later and at different values of deformation recorded and it is strongly related to the maximum value of stress used (Figure 12a–c). This also is represented by no differences between the stages following deformation increasing and decohesion at stress value below the yield point (Figure 12d). It can be explained based on the hardening of the weld at higher values of stress and its small scale at lower ones. Moreover, following these data (Figure 12a) and results from tensile test (Figure 7) the influence of the hourglass measuring zone on the weld response can be evidenced as the stress value exceeding the ultimate tensile strength.

The similarity to conclusions from fracture zones (Figure 11) and displacement courses (Figure 12) total energy versus time (Figure 13) enables to connect its values with stress levels used. Besides, in the case of a high value of stress this parameter increased more rapidly than for the smallest ones, a final value of total energy, representing its huge increase was dependent on the stress levels, taking 5 J (calculated as of 1976 J–1981 J) at 950 MPa (Figure 13a) and two times smaller value at stress, not exceeding the yield point (Figure 13c,d). It can be explained in the same way as in the case of displacement courses i.e., by weld hardening.

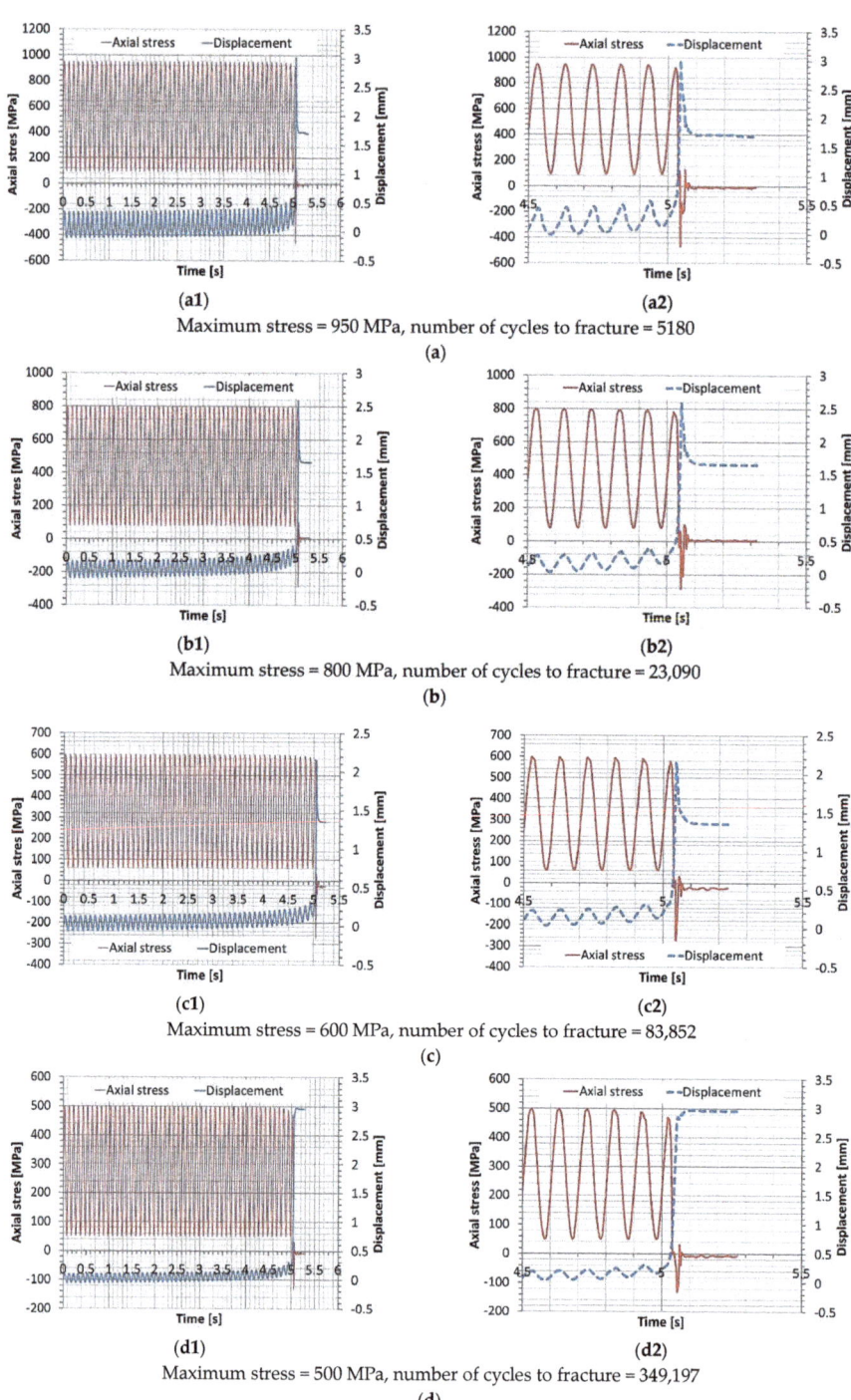

Figure 12. Variations of axial stress and displacement before fracture of the weld under the following values of maximum stress: (**a**) 950 MPa, (**b**) 800 MPa, (**c**) 600 MPa, (**d**) 500 MPa.

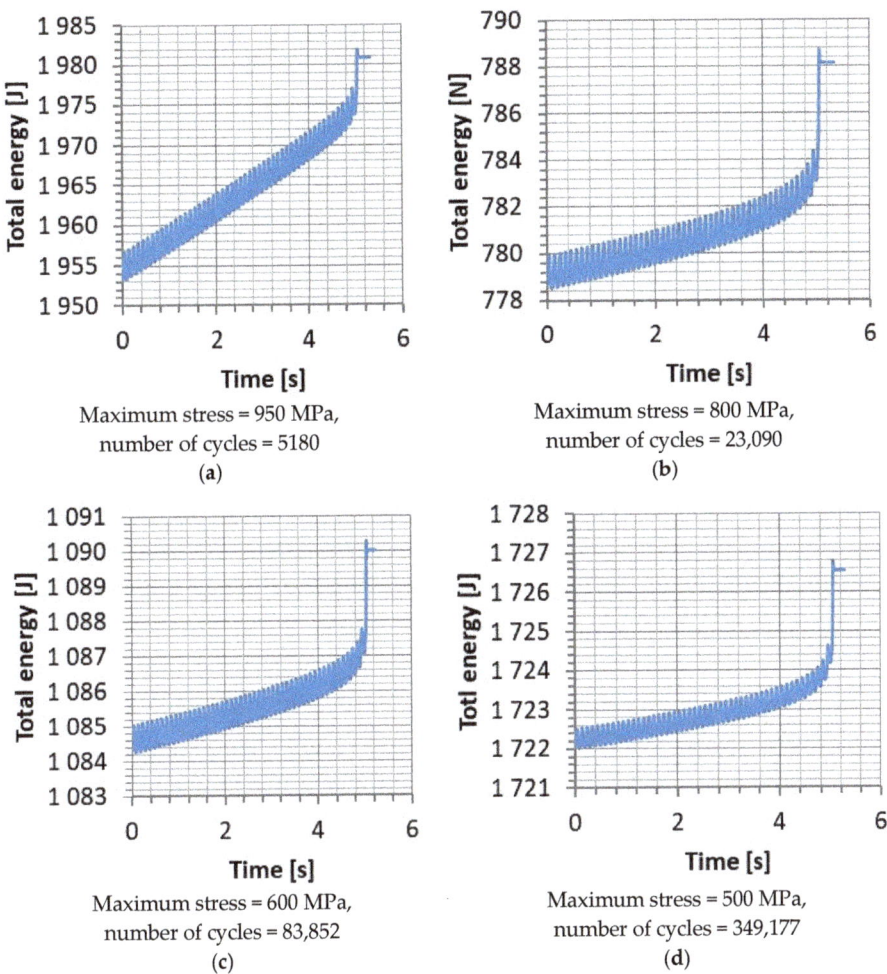

Figure 13. Total energy versus time before fracture of the weld under the following values of maximum stress: (a) 950 MPa, (b) 800 MPa, (c) 600 MPa, (d) 500 MPa.

Generally, differences in the weld behavior under stress level applied were clearly noticed on the relation between axial stress and number of cycles (Figure 14). They were represented by time to fracture containing a few cycles at the highest stress applied 1400 MPa and 2×10^6 for the smallest one i.e., 400 MPa which was established as a fatigue limit of the welded joint because fracture did not occur. Analyzing these data and results from the tensile test, a number of cycles under stress value related to fundamental mechanical properties can be indicated, giving a piece of the knowledge for engineering and researches groups, which use software for designing, calculations, and modelling. They can be directly used for validation of the statistical approach and for elaboration with respect to increasing the prediction accuracy at small-number specimen life data [53].

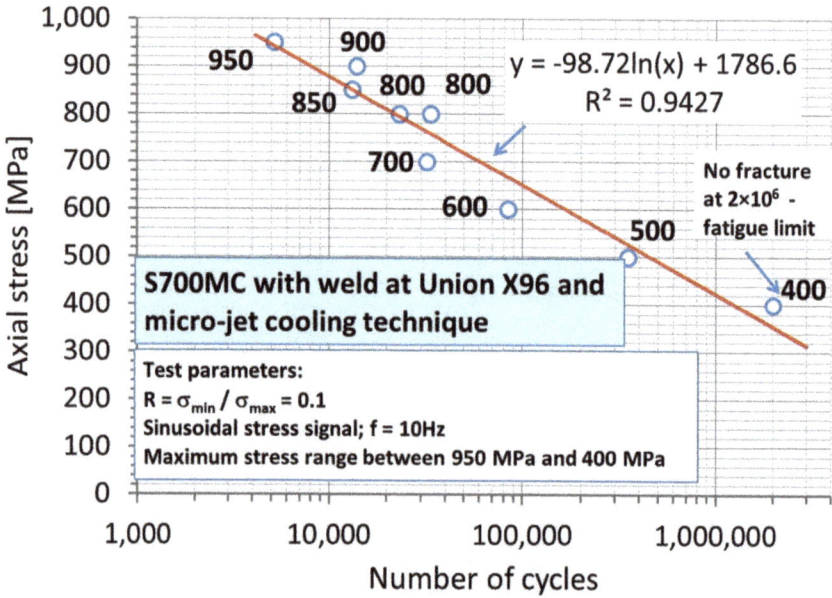

Figure 14. Wöhler curve of S700MC steel welded by means of the welding wire of Union X96.

4. Summary

The article deals with welding of S700MC steel with the use of micro-jet cooling. Welds were made with different electrode wires, various welding parameters and various parameters of micro-jet cooling. After the results of non-destructive tests, main information was obtained on the proper welding parameters, where no cracks appear. Then, the temporary tensile tests were performed and the fatigue experiments were carried out. An influence of the welding process with micro-jet cooling on the steel behavior was represented by 50% reduction of elongation at an almost constant value of ultimate tensile strength. The metallographic structure of the joint was checked. The Wöhler curve was determined and the fatigue limit was presented. These results have enabled to indicate the value of fatigue limit of the micro-jet cooling high strength weld i.e., 400 MPa and its relationship to ultimate tensile strength denoted by 0.48 which supports the efforts of engineers and researches on the elaboration of the welding process and modelling as well as inspection stages. The weld response at initial fatigue cycles exhibited the hardening which directly led to an increase in the stress value used to control the testing machine. This effect disappeared with an increasing number of cycles changing to softening of the weld tested. It was found that, owing to micro-jet cooling, very good mechanical properties of the joint were obtained, as measured by the excellent results of fatigue tests.

5. Conclusions

Micro-jet cooling has been successfully used to weld S700MC steel. It was the first time such a joint was made and its mechanical properties were thoroughly tested. A novelty is also checking the fatigue properties for joints made with using micro-jet cooling.

Based on the research carried out, the following conclusions can be drawn:

1. Good mechanical properties can be obtained when welding S700MC steel by using micro-jet cooling.
2. The properties of the S700MC joint are affected by the selection of thermodynamic welding conditions, including the micro-jet cooling parameters.

3. The conducted non-destructive and destructive tests confirm the correctness of using micro-jet cooling during welding of S700MC steel.
4. The result of the fatigue tests allows us to state that the proposed welding process can be applied to responsible structures.

Author Contributions: Conceptualization, T.W., T.S. and B.S.-L.; methodology, B.S.-L.; software, T.S.; validation, T.W., T.S. and B.S.-L.; formal analysis, B.Ł.; investigation, T.S.; resources, B.S.-L.; data curation, T.W.; writing—original draft preparation, T.S.; writing—review and editing, B.S.-L.; visualization, T.S.; supervision, B.Ł.; project administration, T.W.; funding acquisition, B.S.-L. All authors have read and agreed to the published version of the manuscript.

Funding: This research was funded by (Silesian University of Technology) grant number (BK-205/RT1/2020) and the APC was funded by (Metals (voucher)).

Institutional Review Board Statement: Not applicable.

Informed Consent Statement: Not applicable.

Data Availability Statement: Not applicable.

Acknowledgments: The paper is part of the COST project, CA 18223.

Conflicts of Interest: The authors declare no conflict of interest.

References

1. Lai, C.S.; Jia, Y.; Dong, Z.; Wang, D.; Tao, Y.; Lai, Q.H.; Wong, R.; Zobaa, A.F.; Wu, R.; Lai, L.L. A review of technical standards for smart cities. *Clean Technol.* **2020**, *2*, 290–310. [CrossRef]
2. Connected Vehicles in Smart Cities: The future of transportation. Available online: Interestingengineering.com (accessed on 4 April 2019).
3. Mc McLaren, D.; Agyeman, J. *Sharing Cities: A Case for Truly Smart and Sustainable Cities*; MIT Press: Cambridge, MA, USA, 2015; ISBN 9780262029728.
4. Mohrbacher, H.; Spöttl, M.; Paegle, J. Innovative manufacturing technology enabling light weighting with steel in commercial vehicles. *Adv. Manuf.* **2015**, *3*, 3–18. [CrossRef]
5. Górka, J. Weldability of thermomechanically treated steels having a high yield point. *Arch. Met. Mater.* **2015**, *60*, 469–475. [CrossRef]
6. Lahtinen, T.; Vilaça, P.; Infante, V. Fatigue behavior of MAG welds of thermo-mechanically processed 700MC ultra high strength steel. *Int. J. Fatigue* **2019**, *126*, 62–71. [CrossRef]
7. Suiçmez, A.S.; Piotrowski, M.; Kotyk, M. Fatigue life and type of steel welded joint, steel S650MC and S700MC. *Sci. Tech. J.* **2015**, *5*, 59–68.
8. Krasnowski, K. Possibilities of increasing the fatigue strength of welded joints in steel S700MC through High Frequency Impact Treatment (HiFIT). *Biul. Inst. Spaw.* **2018**, *2018*, 7–15. [CrossRef]
9. Celin, R.; Burja, J. Effect of cooling rates on the weld heat affected zone coarse grain microstructure. *Met. Mater. Eng.* **2018**, *24*, 37–44. [CrossRef]
10. Darabi, J. Development of a chip-integrated micro cooling device. *Microelectron. J.* **2003**, *34*, 1067–1074. [CrossRef]
11. Hashimoto, F.; Lahoti, G. Optimization of set-up conditions for stability of the centerless grinding process. *CIRP Ann.* **2004**, *53*, 271–274. [CrossRef]
12. Muszyński, T.; Mikielewicz, D. Structural optimization of microjet array cooling system. *Appl. Therm. Eng.* **2017**, *123*, 103–110. [CrossRef]
13. Szymczak, T.; Brodecki, A.; Kowalewski, Z.L.; Makowska, K. Tow truck frame made of high strength steel under cyclic loading. *Mater. Today Proc.* **2019**, *12*, 207–212. [CrossRef]
14. Barsukov, V.V.; Tarasiuk, W.; Shapovalov, V.M.; Krupicz, B. Express evaluation method of internal friction parameters in molding material briquettes. *J. Frict. Wear* **2017**, *38*, 71–76. [CrossRef]
15. Bleck, W.; Larour, P.; Baeumer, A. High strain tensile testing of modern car body steels. *Mater. Forum* **2005**, *29*, 21–28.
16. Jaewson, L.; Kamran, A.; Jwo, P. Modeling of failure mode of laser welds in lap-shear specimens of HSLA steel sheets. *Eng. Fract. Mech.* **2011**, *1*, 347–396.
17. Węgrzyn, T.; Szczucka-Lasota, B.; Uscilowska, A.; Stanik, Z.; Piwnik, J. Validation of parameters selection of welding with micro-jet cooling by using method of fundamental solutions. *Eng. Anal. Bound. Elem.* **2019**, *98*, 17–26. [CrossRef]
18. Hadryś, D. Impact Load of Welds after Micro-Jet Cooling / Dynamiczne Obciążenie Spoin Chłodzonych Mikrojetowo. *Arch. Met. Mater.* **2015**, *60*, 2525–2528. [CrossRef]
19. Hadrys, D.; Wegrzyn, T.; Piwnik, J.; Stanik, Z.; Tarasiuk, W. The use of compressed air for micro-jet cooling after MIG welding. *Arch. Metall. Mater.* **2016**, *61*, 1059–1061. [CrossRef]

20. Hobbacher, A. Recommendations for fatigue design of welded joints and components. In *International Institute of Welding Collection*; Springer: Cham, Switzerland, 2016; pp. 11–36.
21. Feng, Z. *Processes and Mechanisms of Welding Residual Stress and Distortion*; Woodhead Publishing Limited and CRC Press LLC. Oak Ridge National Laboratory: Oak Ridge, TN, USA, 2005; p. 364.
22. Totten, G.; Howes, M.; Inoue, T. *Handbook of Residual Stress and Deformation of Steel*; ASM International: Ohio, OH, USA, 2002; p. 499.
23. Sága, M.; Blatnická, M.; Blatnický, M.; Dižo, M.; Gerlici, J. Research of the fatigue life of welded joints of high strength steel S960 QL created using laser and electron beams. *Materials* 2020, *13*, 2539. [CrossRef]
24. Yi, H.; Lee, K.Y.; Bhadeshia, H.K.D.H. Stabilisation of ferrite in hot rolled δ-TRIP steel. *Mater. Sci. Technol.* 2011, *27*, 525–529. [CrossRef]
25. Porter, D.A. Weldable High-Strength Steels: Challenges and Engineering Applications. In Proceedings of the IIW International Conference High-Strength Materials-Challenges and Applications, Helsinki, Finland, 2–3 July 2015.
26. Ma, J.L.; Chan, T.M.; Young, B. Tests on high-strength steel hollow sections: A review. *Proc. Inst. Civ. Eng. Struct. Build.* 2017, *170*, 621–630. [CrossRef]
27. Chatterjee, S.; Murugananth, M.; Bhadeshia, H.K.D.H. δ-TRIP Steel. *Mater. Sci. Technol.* 2007, *23*, 819–827. [CrossRef]
28. Lis, A.K.; Gajda, B. Modelling of the DP and TRIP microstructure in the CMnAlSi automotive steel. *J. Achiev. Mater. Manuf. Eng.* 2006, *15*, 1–2.
29. Goritskii, V.M.; Shneiderov, G.R.; Guseva, I.A. Effect of chemical composition and structure on mechanical properties of high-strength welding steels. *Metallurgist* 2019, *63*, 21–32. [CrossRef]
30. Varelis, G.E.; Papatheocharis, T.; Karamanos, S.A.; Perdikaris, P.C. Structural behavior and design of high-strength steel welded tubular connections under extreme loading. *Mar. Struct.* 2020, *71*, 102701. [CrossRef]
31. Dekys, V.; Kopas, P.; Sapieta, M.; Stevka, O. A detection of deformation mechanisms using infrared thermography and acoustic emission. *Appl. Mech. Mater.* 2014, *474*, 315–320. [CrossRef]
32. Konstrukce. Available online: http://www.konstrukce.cz/clanek/volba-konstrukcnich-oceli-pro-stavebni-svarovane-konstrukce-podle-vyznamu-oznaceni/ (accessed on 17 April 2020).
33. Mansouri, D.; Sendur, P.; Yapici, G.G. Fatigue characteristics of continuous welded rails and the effect of residual stress on fatigue-ratchetting interaction. *Mech. Adv. Mater. Struct.* 2020, *27*, 473–480. [CrossRef]
34. Cremona, C.; Eichler, B.; Johansson, B.; Larsson, T. Improved assessment methods for static and fatigue resistance of old metallic railway bridges. *J. Bridge Eng.* 2013, *18*, 1164–1173. [CrossRef]
35. Kowal, M.; Szala, M. Diagnosis of the microstructural and mechanical properties of over century-old steel railway bridge components. *Eng. Fail. Anal.* 2020, *110*, 10447. [CrossRef]
36. Naib, S.; De Waele, W.; Štefane, P.; Gubeljak, N.; Hertelé, S. Evaluation of slip line theory assumption for integrity assessment of defected welds loaded in tension. *Procedia Struct. Integr.* 2017, *5*, 1417–1424. [CrossRef]
37. Günther, H.-P.; Hildebrand, J.; Rasche, C.; Versch, C.; Wudtke, I.; Kuhlmann, U.; Vormwald, M.; Werner, F. Welded connections of high-strength steels for the building industry. *Riv. Ital. Saldatura* 2014, *66*, 1055–1087. [CrossRef]
38. Valicek, J.; Czan, A.; Harnicarova, M.; Sajgalik, M.; Kusnerova, M.; Czanova, T.; Kopal, I.; Gombar, M.; Kmec, J.; Safar, M. A new way of identifying, predicting and regulating residual stress after chip-forming machining. *Int. J. Mech. Sci.* 2019, *155*, 343–359. [CrossRef]
39. Fatigue Analysis on the Web. Available online: https://www.efatigue.com/ (accessed on 20 April 2020).
40. *Metallic Materials—Tensile Testing-Part 1: Method of Test at Room Temperature*; Standard ISO 6892-1; Beuth-Verlag: Berlin, Germany, 2019.
41. Cheng, X.; Fischer, J.W.; Prask, H.J.; Gnäupel-Herold, T.; Yen, B.T.; Roy, S. Residual stress modification bypost-weld treatment and its beneficial effect on fatigue strength welded structure. *Intern. J. Fatigue* 2003, *25*, 1259–1269. [CrossRef]
42. Stephens, R.I.; Fatemi, A.; Stephens, R.R.; Fuchs, H.O. *Metal Fatigue in Engineering*; Wiley Interscience: New York, NY, USA, 2001, p. 496.
43. Lacalle, R.; Álvarez, L.; Ferreño, D.; Portilla, J.; Ruiz, E.; Arroyo, B.; Gutiérrez-Solana, F. Influence of the flame straightening process on microstructural, mechanical and fracture properties of S235 JR, S460 ML and S690 QL structural steel. *Exp. Mech.* 2013, *53*, 893–909. [CrossRef]
44. Górka, J. Assessment of the weldability of T-welded joints in 10 mm Thick TMCP steel using laser beam. *Materials* 2018, *11*, 1192. [CrossRef] [PubMed]
45. *Strenx 700MC D/E.*; Data sheet 2008; SSAB: Stockholm, Sweden, 20 April 2017. Available online: https://www.ssab.com/products/brands/strenx/products/strenx-700-mc (accessed on 15 October 2020).
46. Metallic Materials. *Tensile Testing—Method of Test at Room Temperature*; PN-EN ISO 6892-1:2020; ISO: Warsaw, Poland, 2020.
47. *Standard Practice for Presentation of Constant Amplitude Fatigue Test Results for Metallic Materials*; ASTM E468-18; ASTM International: West Conshohocken, PA, USA, 2018.
48. *Standard Practice for Conducting Force Controlled Constant Amplitude Axial Fatigue Tests of Metallic Materials*, ASTM E466-15; ASTM International: West Conshohocken, PA, USA, 2015.
49. Araque, O.; de la Peña, N.A.; Laguna, E.H. The effect of weld reinforcement and post-welding cooling cycles on fatigue strength of butt-welded joints under cyclic tensile loading. *Materials* 2018, *11*, 594. [CrossRef]

Song, W.; Liu, X.; Berto, F.; Razavi, S.M.J. Low-cycle fatigue behavior of 10crni3mov high strength steel and its undermatched welds. *Materials* **2018**, *11*, 661. [CrossRef]

Mutombo, K.; Du, M. Corrosion fatigue behaviour of aluminium 5083-h111 welded using gas metal arc welding method. *Arc Welding* **2011**, *139*, 105789. [CrossRef]

Li, C.; Wu, S.; Zhang, J.; Xie, L.; Zhang, Y. Determination of the fatigue P-S-N curves—A critical review and improved backward statistical inference method. *Int. J. Fatigue* **2020**, *139*, 105789. [CrossRef]

Welding—Fusion-Welded Joints in Steel, Nickel, Titanium and their Alloys (Beam Welding Excluded)—Quality Levels for Imperfections; PN-EN ISO 5817:2014; ISO: Warsaw, Poland, 2014.

Recrystallization Behavior of a Pure Cu Connection Interface with Ultrasonic Welding

Zhanzhan Su [1], Zhengqiang Zhu [1,*], Yifu Zhang [1,2], Hua Zhang [1] and Qiankun Xiao [1]

[1] Key Laboratory for Robot and Welding Automation of Jiangxi Province, School of Mechanical and Electrical Engineering, Nanchang University, Nanchang 330031, China; suzhanzhan@email.ncu.edu.cn (Z.S.); zhangyifu@jju.edu.cn (Y.Z.); hzhang@email.ncu.edu.cn (H.Z.); 355906316001@email.ncu.edu.cn (Q.X.)
[2] School of Mechanics and Materials Engineering, Jiujiang University, Jiujiang 332005, China
* Correspondence: zhuzhq@ncu.edu.cn; Tel.: +86-186-7985-8610

Abstract: Three-dimensional metal waveguide components are key components in the next generation of radio telescopes. Ultrasonic additive manufacturing technology combining ultrasonic welding and micro electrical discharge machining (micro-EDM) provides a new method for the overall manufacturing of waveguide elements, and the effective welding of Electrolytic Tough Pitch copper (Cu-ETP) sheets is the key process of this method. This study demonstrates that the orthogonal test optimization method is used to conduct ultrasonic welding tests on Cu-ETP. Specifically, electron backscattered diffraction (EBSD) technology is used to analyze the crystal grains, grain boundary types and texture changes during interface recrystallization. In addition, the finite element software ABAQUS 6.13 is employed to calculate the temperature field in order to determine the possibility of recrystallization of the welding interface. The results showed that the average grain size of the welding interface decreased from 20 to 1~2 µm. The Cu-ETP matrix is mainly composed of coarse grains with high-angle grain boundaries (HAGBs), while a large number of low-angle grain boundaries (LAGBs), subcrystals and fine equiaxed grains appear in the welded joint. At the same time, discontinuous dynamic recrystallization (DDRX) occurs in the less strained area, and continuous dynamic recrystallization (CDRX) is predominant in the greater strain area. The temperature field calculation shows that the peak temperature of the welding interface exceeds the recrystallization temperature of Cu-ETP from 379.05 to 433.2 °C.

Keywords: three-dimensional metal waveguide components; ultrasonic welding; electron backscattered diffraction (EBSD); finite element method; recrystallization

1. Introduction

Metal waveguide components (with a three-dimensional distributed waveguide network, including receiving signal holes, local oscillator signal holes, coupling holes, etc.) are an important part of integrated heterodyne array receivers in the sub-millimeter wave band of radio telescopes. In the past, discrete waveguide components were connected by pin positioning screws to join them together, making it difficult to achieve large-scale pixels. The next-generation of array receivers with hundred-beam scale pixels must be integrated and modular. The Purple Mountain Observatory of China has proposed a 3D waveguide network with vertically distributed receiving signal holes. The 3D waveguide network must be formed as a whole, and the structure must be very compact [1]. The key to realizing this array pattern is high-precision 3D waveguide network manufacturing. Therefore, the integral forming of three-dimensional metal waveguide components is a key link in the preparation of next-generation radio telescopes [2]. A schematic diagram of the 3D metal waveguide network is shown in Figure 1. The machining accuracy of the 3D waveguide network coupling channel is controlled within 0.01 mm. The current high-speed milling technology in traditional ultrasonic additive manufacturing cannot process such coarse holes [3–7]. However, micro electrical discharge machining (micro-EDM) technology ensures the shape accuracy, dimensional accuracy, and surface quality of the micron-level 3D waveguide network and can form sharp corners inside the component [8]. Figure 2 shows the shape and size of the straight groove for the micro-EDM. It can be seen that the processed straight groove has a good size consistency, the error is controlled at approximately 5 µm, and the surface roughness reaches 0.6 µmRa. The new ultrasonic additive manufacturing technology integrating ultrasonic welding and micro-EDM provides a new method for the overall preparation of waveguide elements.

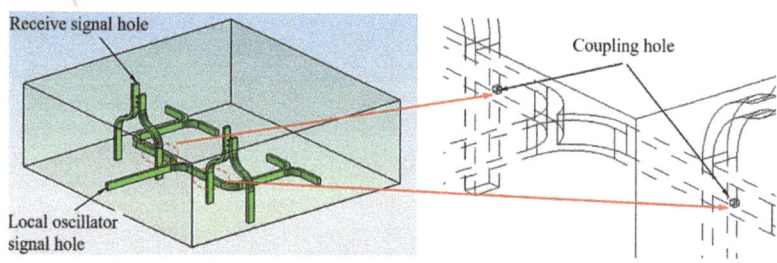

Figure 1. Schematic diagram of the 3D metal waveguide network.

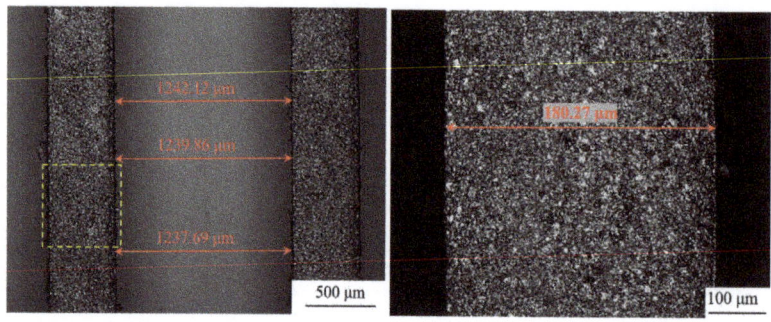

Figure 2. Straight groove shape and size of the micro-EDM.

The waveguide network is usually made of metallic materials (copper, aluminum, etc.). Considering that the receiver will work at extremely low temperatures (as low as 4 K), the thermal conductivity of aluminum alloys will deteriorate at low temperatures. Therefore, Electrolytic Tough Pitch copper (Cu-ETP) is used as the research material. The effective connection of the Cu-ETP sheet is the key process for realizing the application of this technology. Ultrasonic welding technology only needs 25–50% of the melting point of the metal to realize the metallurgical connection between metal layers through ultrasonic friction, avoiding the liquid–solid phase transition, residual stress, dimensional changes and metallurgical incompatibility. The technical principle is shown in Figure 3. The vibration direction is different from that of plastic ultrasonic welding. Longitudinal vibration with pressure causes the metal sheet (the thickness is generally 0.3 mm) to achieve metallurgical bonding [9–14]. Fujii, H.T. et al. [15] studied the formation mechanism and joint failure mode of ultrasonic welding. The results show that metal bonding is the main connection mechanism of Al–Al, and physical bonding caused by friction is the main connection mechanism of the Cu–Cu or Al–Cu interface. Joint failure is a mixed failure mode of weld interface peeling and joint tearing. Nambu, S. et al. [16] found that high-frequency tangential motion during the ultrasonic welding process caused the metal interface to form nanoparticles. Under the action of the welding pressure, the nanoparticles are compacted to form a bonding interface, which is formed by mixing two specimen components. Sriraman, M.R. et al. [17] realized the high-power ultrasonic additive manufacturing of copper foil with a thickness of 150 µm at room temperature.

The results show that copper involves significant material softening and plastic flow in the ultrasonic additive process. The initial grains of copper foil (25 µm in size) transform into fine dynamic recrystallized grains (0.3–10 µm in size) at the weld interface; this phenomenon is believed to be metallurgical bonding caused by a grain boundary migration. Xie, J. et al. [18] studied the recrystallization behavior of aluminum alloy ultrasonic welds by comparing the microstructure and texture of the base material and the welded specimen. The results showed that dynamic recovery (DRV) makes the weld form a layered structure with {111} and rotating cube {001} shear textures. Dynamic recrystallization (DRX) causes the weld to form uneven fine/ultrafine equiaxed grains, and the weld area has a sharp recrystallized structure {311}. We systematically explored the DRV and DRX mechanisms that control the evolution of the microstructure during the ultrasonic welding process. Fujii, H.T. et al. [19] studied the microstructure and mechanical properties of aluminum alloy/stainless steel ultrasonic welded joints. The results show that with an increasing welding energy, the fracture mode of the lap joints changes from interface peeling to base metal fracture during tension. During the welding process, the weld produces a high strain rate shear deformation and temperature rise. Through the recrystallization of aluminum alloy, the microstructure of the weld is composed of fine grains and equiaxed grains. Haddadi, F. et al. [20] used electron backscattered diffraction (EBSD) technology to study the dynamic recrystallization and grain growth of the weld during the ultrasonic welding of an aluminum alloy. The results show that high-power ultrasonic welding causes large-scale plastic deformation in the welded area. The welding seam temperature increased to 440 °C within a short welding time of 100 ms, and an ultrafine grain structure was observed at the welding interface. With an increasing welding time, the texture changed from an initially cube-dominated texture, to one where the typical β-fiber brass component prevailed.

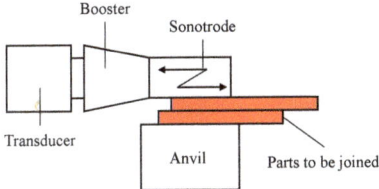

Figure 3. Scheme of metal ultrasonic welding.

Due to the interaction of complex factors in the ultrasonic welding process, the detailed evolution of the recrystallization behavior of the weld interface is still unclear. Moreover, the current research focuses on understanding the recrystallization behavior of metals away from the interface, and the two main bonding mechanisms on the interface are not well understood. In addition, previous work was limited to aluminum alloys, which led to insufficient research on the ultrasonic welding of copper for waveguides. In this paper, the mechanical properties (tensile load, peeling load) and the microstructure (metallography, EBSD) of welded joints are studied to evaluate the welding quality of welded joints with various process parameters, in order to study the connection mechanism of ultrasonic welding and to obtain the optimal welding parameters. EBSD technology is used to analyze the grain changes, grain boundary type changes, texture changes, etc., during the recrystallization process of the copper welding interface. ABAQUS finite element software is used to simulate the temperature field of the ultrasonic welding process of copper, and the temperature field is used to judge whether the weld temperature reaches the recrystallization temperature. The systematic exploration of this work is helpful at explaining the connection mechanism of ultrasonic welding and has important guiding significance for the control of the weld's microstructure. The research results will provide guidance for the design process of manufacturing 3D metal waveguide devices using new ultrasonic additive manufacturing technology.

2. Experimental and Finite Element Model

2.1. Experimental Materials and Methods

In this paper, Cu-ETP was used for the welding experiment of lap joints. The specimen size was 100 mm × 20 mm × 0.3 mm, and the overlap amount was 30 mm. The model for the ultrasonic welding equipment used in the test was NC-2020A (system power $P = 50 \sim 2000$ W, system working force $F = 460 \sim 1400$ N, welding time $t = 0 \sim 10^4$ ms, welding spot area 15 × 6 mm², ultrasonic vibration frequency $f = 20$ kHz). To evaluate the relationship between the ultrasonic welding process parameters and the predicted response value (welded joint tensile and shear load), the experiment adopted an orthogonal experimental design and conducted a range analysis to determine the optimal process parameter combination [21,22].

The factor level codes used in the ultrasonic welding experiment are shown in Table . Among them, "0" is the center point, "−" is the lower level, and "+" is the upper level.

Table 1. Level codes and true values of the ultrasonic welding process parameters.

Level Coding	Factor		
	Welding Power, P (W)	Welding Time, t (ms)	Welding Force, F (N)
−	1200	500	600
0	1400	700	900
+	1600	900	1200

Before welding, 320#SiC sandpaper was used to remove the surface oxide of the base metal, and acetone was finally used to completely degrade the surface oxide. The weld cross-section specimens were prepared by wire cutting. The corrosive agent $FeCl_3:HCl:H_2O$ = 3 g:10 mL:100 mL was used to corrode the weld, and the corrosion time was 3–5 s. An Axio Imager A1m optical microscope was used to observe the metallographic structure of the joint. The evolution of the microtexture of the joint was analyzed by EBSD (Oxford technology. During the test, the specimen tilt angle was 70°, the working distance was approximately 20 mm, and the acceleration voltage was 20 kV. The test data were analyzed using Aztec and Channel 5 software. A WDW3050 universal testing machine (Jinan Time Shijin Testing Machine Co., Ltd., Jinan, Shandong Province, China) was used to test the tensile load and peel load of the specimen, and the loading speeds were 3 and 15 mm/min respectively. HVS-1000 micro Vickers hardness tester (Jiangxi Huake Precision Instrument Co., Ltd., Nanchang, Jiangxi Province, China) was used to measure the microhardness of the joint, the load was 0.025 kg, and the pressure holding time was 10 s.

To avoid systematic errors, a random ultrasonic welding experimental design is performed. The process parameters and response results are shown in Table 2.

Table 2. Orthogonal table of the copper ultrasonic welding experiment.

Experiment No.	Mode	P (W)	t (ms)	F (N)	Tensile Force, F_τ (N)
1	− − −	1200	500	600	1707.0
2	− −0	1200	700	900	1780.5
3	− −++	1200	900	1200	1819.1
4	0−0	1400	500	900	1852.4
5	00+	1400	700	1200	1923.4
6	0+−	1400	900	600	1887.7
7	+−+	1600	500	1200	1854.4
8	+0−	1600	700	600	1825.8
9	++0	1600	900	900	1886.8

2.2. Finite Element Model

Due to the symmetry of the welded structure geometry, the boundary constraints and the load in the longitudinal plane, the finite element model was simplified based on symmetry to reduce the amount of calculation. It was assumed that there was no gap and no relative sliding on the contact surface of the sonotrode, Cu-ETP and anvil. The room temperature was 20 °C, and the sonotrode area was equal to the plastic deformation zone A_{DZ}. ABAQUS finite element software was used to establish the ultrasonic welding model, as shown in Figure 4. The physical property parameters of the sonotrode, anvil and Cu-ETP are shown in Tables 3 and 4 [23].

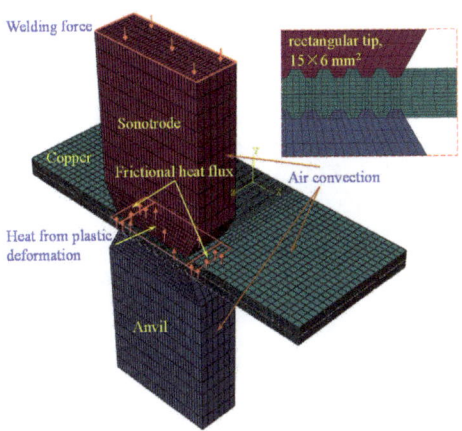

Figure 4. Finite element model of Cu-ETP ultrasonic welding.

Table 3. Physical property parameters of the sonotrode and anvil.

Temperature (°C)	Elastic Modulus (GPa)	Density (kg/m³)	Poisson's Ratio	Linear Expansion Coefficient (10^{-5}/°C)	Thermal Conductivity (W/m·K)	Specific Heat Capacity (J/kg·K)
20	212	7800	0.29	1.48	34	460
300	192	7800	0.29	1.51	36.2	513
500	156	7800	0.29	1.58	38.1	532
800	139	7800	0.29	1.65	39.5	560
1200	107	7800	0.29	1.72	40.7	584
1500	83	7800	0.29	1.75	41.2	607

Table 4. Physical property parameters of Cu-ETP.

Temperature (°C)	Elastic Modulus (GPa)	Density (kg/m³)	Yield Strength (MPa)	Poisson's Ratio	Linear Expansion Coefficient (10^{-5}/°C)	Thermal Conductivity (W/m·K)	Specific Heat Capacity (J/kg·K)
20	109	8969	804	0.35	1.64	390	385.2
100	105	8969	620	0.35	1.68	382	406
200	102	8969	638	0.35	1.72	371	406
300	98	8969	552	0.35	1.75	352	410
400	92	8969	455	0.35	1.79	348	412
500	90	8969	377	0.35	1.79	343	416
600	87	8969	303	0.35	1.81	339	418

The heat source of ultrasonic welding mainly comes from the heat generated by the plastic deformation of the workpiece and the interface friction [24]. The heat flow generated due to plastic deformation was applied to the plastic deformation zone A_{DZ} below the sonotrode. The heat flow generated by the ultrasonic vibration friction of the welding interface was applied to the interface friction area A_{FR} near the sonotrode. The heat flow equation is calculated as follows:

The heat flow equation is generated by plastic deformation.

$$Q_w = \frac{P_{TOTAL}}{A_{DZ}} = \frac{F_w(T, F_N, t) \cdot V_{avg}}{A_{DZ}} \quad (1)$$

where

$$F_w = (T, F_N, t) = \sqrt{[\frac{Y(T)}{2}]^2 - (\frac{F_N/A_{DZ}}{2})} \cdot A_w(t) \quad (2)$$

$$V_{avg} = 4\varepsilon_0 f_w \quad (3)$$

where P_{TOTAL} is the welding power; A_{DZ} is the plastic deformation area; F_w (T, F_N, t) is the welding force [25]; V_{avg} is the ultrasonic welding average speed; $Y(T)$ is the workpiece yield strength; F_N is the welding static pressure; $A_w(t)$ is the welding area; ε_0 is the ultrasonic amplitude; and f_w is the vibration frequency.

In a short welding time, the interface only produces a small area of mechanical interlocking, and the connection area is much smaller than the end face of the sonotrode. The welding area can be approximated as:

$$A_w(t) = 1.6 \cdot \left(1 - e^{-kt/sec}\right) \cdot 10^{-5} \cdot m^2 \text{ (k is a constant)} \quad (4)$$

The heat flow equation generated by interfacial friction:

$$Q_{FR} = \frac{P_{FR}}{A_{FR}} = \frac{F_{FR} V_{avg}}{A_{FR}} \quad (5)$$

$$F_{FR} = \mu \cdot F_N \quad (6)$$

$$Q_{FR} = \frac{4\mu F_N \varepsilon_0 f_w}{A_{FR}} \quad (7)$$

where P_{FR} is the frictional heat generation power; A_{FR} is the interface friction area; F_{FR} is the friction force; and μ is the friction coefficient.

3. Results and Discussion

3.1. Effect of the Process Parameters on the Response Value

From the results of the variance analysis in Table 5, it can be seen that $R^2 > 90\%$, and the probability $\Delta Prob > \Delta F$ is less than 0.05, indicating that the model can simulate the real process. Figure 5 is a scatter diagram of the predicted response and its actual values, indicating that the regression model is not unfit. From the comparison of $\Delta Prob > \Delta F$ values, it is determined that the degree of influence of each parameter on the tensile load is $P > t > F$. This paper adopts general optimization standards for copper ultrasonic welding and sets the target as the maximum ultimate tensile strength. Additionally, the welding process parameters are kept within the design space domain. From the main effect diagram of the signal to noise (SN) ratios in Figure 6, the optimal process parameters combination can be predicted as $P = 1400$ w, $t = 700$ ms, and $F = 1200$ N.

The influence of various factors on the response value of Cu-ETP ultrasonic welded joints includes the welding energy P, welding time t and welding force F, and the influence of the various process parameters on joint quality is not simply superimposed but instead interactive. Therefore, the influence of the welding heat input ($P \times t$) and welding force F on joint quality is analyzed below.

Table 5. Analysis of the variance results.

	df	Sum of Squares	Mean Square	F Value	ΔProb > ΔF
Intercept	6	33,473.28	5578.88	94.27	0.0105
P	2	22,722.00	11,361.00	191.97	0.0052
t	2	5538.23	2769.11	46.79	0.0209
F	2	5213.05	2606.52	44.04	0.0222
Residual	2	118.36	59.18		
Cor Total	8	33,591.64			

R-Squared—99.65%; Adj R-Squared—98.59%; Pred R-Squared—92.86%; Adeq Precision—31.896.

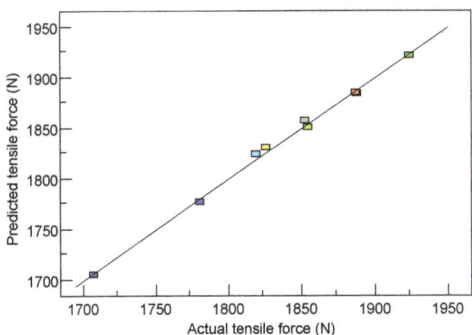

Figure 5. Scatter diagram of the predicted and actual tensile force.

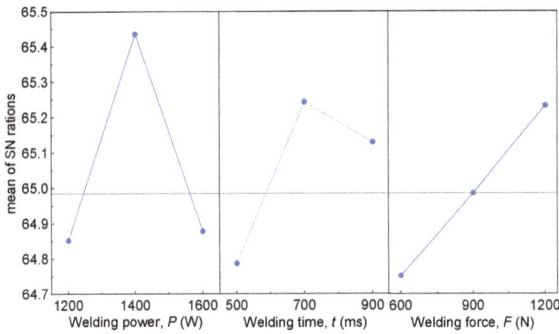

Figure 6. Main effects plot for SN rations.

Figure 7 shows the metallographic structure of Cu-ETP welds at different welding heat inputs when the welding force is 1200 N. At a low welding heat input, the quality of the joint is only determined by the welding force and the ultrasonic vibration to achieve mechanical fitting of the local contact surface. With an increase in the welding heat input, the welding workpiece appears softened, and a small amount of plastic deformation makes the interface tightly connected. When the heat input is 700 ms × 1400 W, the connection line is not obvious, which also shows that the joint has the best mechanical performance. With a further increase in heat input, the welding workpiece softens, and the plastic deformation intensifies, resulting in a stress concentration in the local connection area and a decrease in the mechanical properties. The corresponding tensile load and peeling load curves are shown in Figure 8.

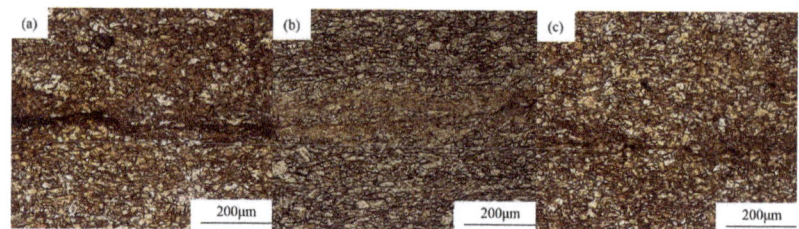

Figure 7. Metallographic diagram of the joints with different welding heat inputs: (**a**) 700 J; (**b**) 980 (**c**) 1260 J.

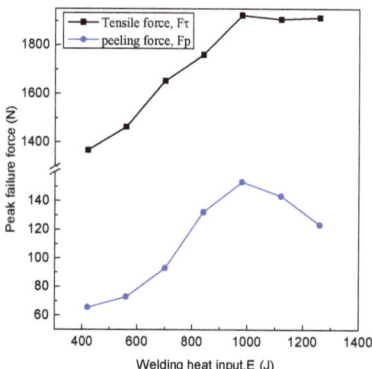

Figure 8. Tensile and peeling peak failure load with different welding heat inputs.

Figure 9 shows the metallographic structure of Cu-ETP welds at different welding forces when the welding heat input is 500 ms × 1400 w. It can be seen that due to the low static friction of the interface under a low welding force, false solder appears on the bonding surface. When the welding force is 1200 N, the grains near the weld are significantly refined and the joint is well bonded. Excessive welding force causes severe deformation of the weld and microcracks in some areas, which reduces the quality of the joint connection. The corresponding tensile load and peel load curves are shown in Figure 10.

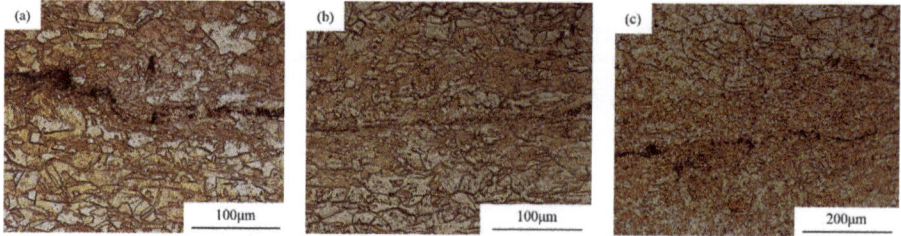

Figure 9. Metallographic diagram of the joints with different welding forces: (**a**) 600 N; (**b**) 1200 N; (**c**) 1400 N.

The ultrasonic welding process is complicated, involving not only mechanical interlocking but also interatomic diffusion and dynamic recrystallization of crystal grains [26]. In the early stage of Cu-ETP ultrasonic welding, the welds were not directly bonded but instead partially mechanically interlocked. Due to the low welding heat input or low welding force, the welding energy between the interfaces is insufficient. Under the action of an external force, the connecting surfaces squeeze and cause friction on one another to form a mechanical interlock. Mechanical interlocks are usually jagged. This type c interlocking is only macroscopic mechanical interlocking, and it is easy to detach under the action of an external force, so the tensile load and peeling load of this joint are relatively low. As the welding energy or welding force increase, the mechanical interlock will gradually decrease. This shows that mechanical interlocking is not the main connection mechanism of copper ultrasonic welding.

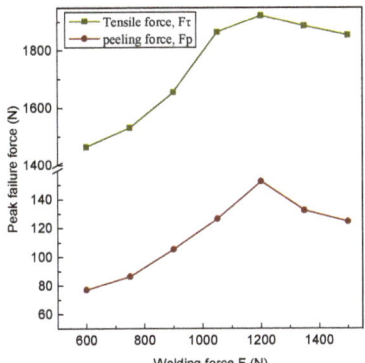

Figure 10. Tensile and peeling peak failure load with different welding forces.

Under the action of static pressure and high-frequency vibration, the interface temperature of the joint increases rapidly. When the energy acquired by the atom is greater than its activation energy for diffusion, the atom diffuses, and the metal atoms interact to form a bond. Metal bonding enhances the mechanical properties of welded joints. This shows that metal bonding plays a key role in the ultrasonic welding of Cu-ETP. The grain boundary misorientation angle is less than 5° as indicated by the green line in Figure 11, and the grain boundary misorientation angle is greater than 5°, as marked by the black line in Figure 11. Figure 11a,b shows that the quantity of small-angle grain boundaries increases significantly after welding. The fraction of grain boundary misorientation distribution in Figure 11c,d also shows this trend. These results lie in the plastic deformation of the welded joint during the welding process, which leads to an increase in the dislocation density. The welding process increases the temperature of the substrate, which leads to the climbing of ductile dislocations and the cross-slip of screw dislocations. The dislocations gradually evolve into multilateral structures and then develop into small-angle grain boundaries. From the grain distribution of Cu-ETP welded joints with the optimal process, it shows that there are a large number of coarse columnar grains far away from the weld area. Due to recrystallization, a large number of small equiaxed grains with a diameter of a few microns are generated in the area adjacent to the weld. These newly generated fine equiaxed grains increase the number of grain boundaries, hinder the movement of dislocations and improve the strength and plasticity of the material. Figure 11e shows the distribution of the micro Vickers hardness of the joint. It can be seen that the microhardness of the connecting interface area is higher than that of the base material (~91 HV). These results further indicate that severe plastic deformation, grain refinement and dislocation multiplication occurred in the interface zone.

3.2. Analysis of Interface Recrystallization Behavior

The microstructure of Cu-ETP mainly includes coarse grains with HAGBs (phase difference $\alpha > 15°$), with a large number of annealing twins. The base metal structure is unevenly distributed, and the grain size is 10~30 µm (Figure 12a). The misorientation distribution of the base material grains (columnar line) is essentially consistent with the random distribution curve (black curve). This shows that the misorientation distribution of the copper base material obtained by annealing conforms to the theoretical value. In addition, these misorientations include not only the grain boundaries, but also the sub-grain boundaries inside the grains (Figure 12b).

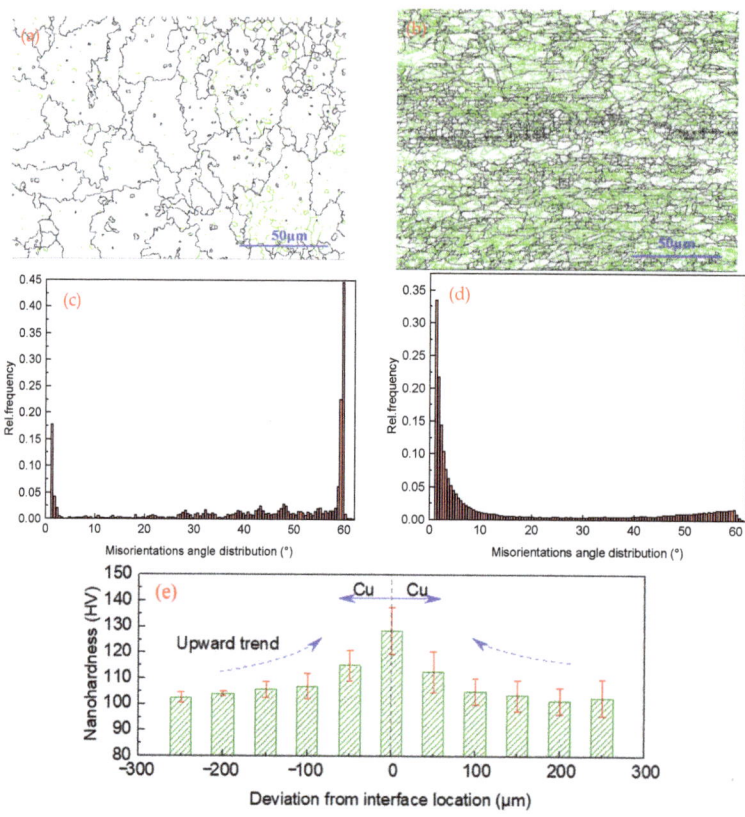

Figure 11. Grain boundary misorientation distribution map: (**a**) Cu-ETP before welding; (**b**) the welded joint at optimal process parameters combination. Fraction of grain boundary misorientation distribution: (**c**) Cu-ETP before welding; (**d**) the welded joint at optimal process parameters combination. (**e**) Nanohardness histogram of the joint.

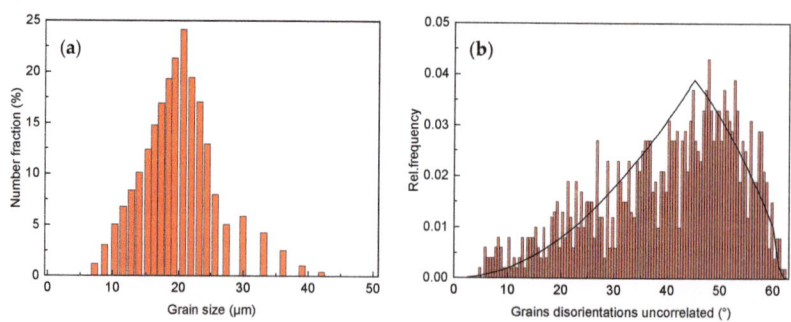

Figure 12. Statistical graph of the base metal grain size (**a**) and misorientation (**b**).

The grain boundary map of the copper ultrasonic welded joint is shown in Figure 13. The purple line represents the high-angle grain boundary. It can be seen that there are no visible defects in the weld, such as cracks or virtual welds. The area far away from the welding interface is composed of coarse deformed grains and a large number of low-angle grain boundaries (LAGBs; phase difference α < 15°). These LAGBs are generally accompanied by the generation of subcrystals and dislocations. The gradual disappearance of copper twins is due to the welding pressure and ultrasonic vibration that promote the deformation and movement of the twins, thereby forming coarse deformed grains. The welding interface area is mainly composed of uneven regenerated small equiaxed grains, which are gathered together in large numbers with a minimum size of 0.5 μm. This is because the fine grains regenerated during the ultrasonic welding process aggregate under the action of plastic flow. When the temperature reaches the recrystallization temperature, the original crystal defects (dislocations, grain boundaries, etc.) will re-nucleate and grow into distortion-less equiaxed crystals. Therefore, the appearance of small equiaxed crystals in the joint is an important sign of copper recrystallization during ultrasonic welding. In addition, a small number of slender layered structures are generated in the interface area, which is related to the dynamic recovery effect during the ultrasonic welding process. At the same time, a large number of HAGBs are generated, while the LAGBs are significantly reduced, which is due to the copper recrystallization temperature reached during the welding process. Some LAGBs (subcrystals) merge with adjacent subcrystals to form high-angle grain boundaries, and the other LAGBs (dislocations) are entangled and migrate to form high-angle grain boundaries due to an increase in the external pressure and activation energy.

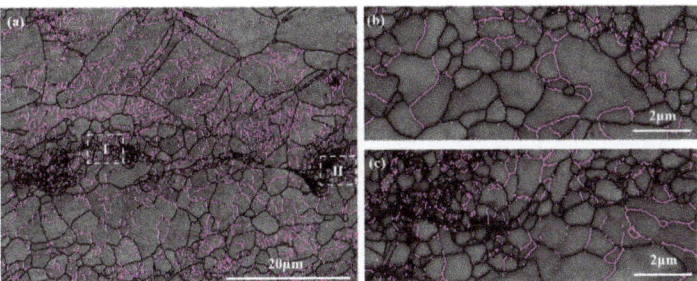

Figure 13. Grain boundary map of the Cu-ETP ultrasonic welded joint (**a**), grain boundary map in zone I (**b**) and II (**c**).

Figure 14 is a statistical graph of the grain size and misorientation of the welded joint. The grain size of copper is obviously reduced after ultrasonic welding, and the average grain size is 1~2 μm. The grain size is relatively concentrated, and most of the grain size is in the range of 1~10 μm, which shows that ultrasonic welding has the effect of grain refinement (Figure 14a). It can be seen from the misorientation statistical graph that the misorientation has changed from high to low. This shows that the small-angle grain boundaries increase after ultrasonic welding, and the large-angle grain boundaries decrease, which corresponds to Figure 13 (Figure 14b).

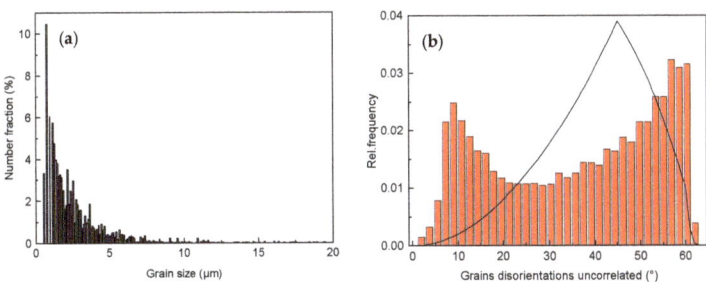

Figure 14. Statistical graph of the welded joint grain size (**a**) and misorientation (**b**).

There is no strong oriented texture in the copper base material, mainly brass texture {112}<111>, (<001>‖ND) and a cubic texture {001}<100>, (<001>‖ND). Figure 15 shows the distribution of the EBSD inverse pole figure (IPF) of the Cu-ETP ultrasonic welding interface microstructure and the crystal plane pole figure (PF) of the microstructure as {100}, {101} and {001}. It can be seen that the welds mainly have shear textures {111}<143>, {111}<110> (blue position) and {221}<122> (purple position), and parts of the {001}<130>, {001}<230> and {114}<131> textures, etc. Among them, there are many regenerated fine cubic textures {100}<001> (red position) and shear textures {111}<110> at the welding interface. The evolution of the interface texture is related to the DRV and DRX mechanism during ultrasonic welding.

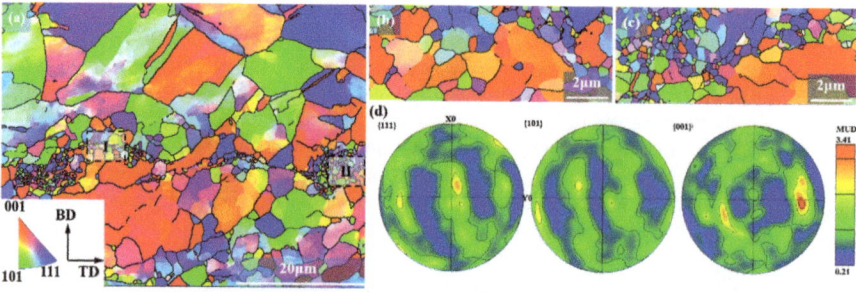

Figure 15. EBSD maps with the corresponding pole figures: (**a**) IPF, IPFs in zone I (**b**) and II (**c**), (**d**) PFs.

The orientation distribution function (ODF) diagram can be used to quantitatively and accurately analyze the spatial distribution of the grain orientation in the texture material to further characterize the microstructure of the weld (Figure 16). The ODF map was measured to obtain the Euler angle, and the Euler angle was used to calculate the micro-textures in different regions. It can be seen that there are {001} <100> cubic textures {001} <110> rotating cube textures and {111} <143>, {221} <122> deformation textures in the Cu-ETP welds. This cubic texture is not the cubic texture of the base material but is instead produced during the recrystallization process with ultrasonic welding. The cubic texture in the base metal becomes the core of recrystallization under the action of ultrasonic excitation and welding pressure and finally grows into a new recrystallized cubic texture. The cubic texture is also one of the main recrystallization textures of face-centered cubic crystals. The deformation texture is produced by the plastic deformation of the joint during the welding process. The reason is that the base material grains are produced via plastic deformation, and the newly formed recrystallized grains are deformed during the ultrasonic welding process.

The recrystallization behavior of metallic materials occurs within a certain temperature range. The recrystallization temperature is not a fixed physical constant but is instead affected by many factors. The most significant factor is the degree of metal deformation. As the degree of deformation increases, the more energy stored inside the material, the greater the driving force for recrystallization, which reduces the recrystallization temperature. The grain size of the base material also has a greater influence on the recrystallization temperature. The smaller the grain size is, the lower the recrystallization temperature. This is because most of the recrystallization nuclei are located at grain boundaries and dislocations. The finer the grains, the more grain boundaries there are, and the easier recrystallization and nucleation are. In addition, there are other factors that affect the recrystallization temperature, such as material solute atoms and second-phase particles. Generally, in pure metals with a large degree of deformation, the recrystallization temperature is approximately (0.35–0.4) Tm (the melting point of pure metals). The melting point of copper is 1083°C, so the recrystallization temperature of copper is approximately 379.05–433.2 °C.

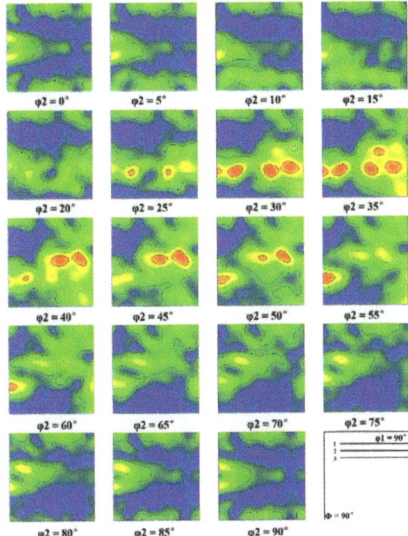

Figure 16. ODF diagram of a copper ultrasonic welding joint.

3.3. Numerical Simulation Results and Analysis

Figure 17 shows the simulation results of the temperature field with different welding heat inputs at a welding force of 1200 N. As the welding heat input increases, the temperature at the center of the welding interface also increases. The center point temperatures are 573, 723 and 858 °C, while the melting point of copper is 1083 °C. The temperature below the melting point indicates that ultrasonic welding is a solid phase connection technology. The maximum temperatures reached 52.9%, 66.8% and 79.2% of the melting point of copper, respectively. Within the range of 30% to 80%, the simulation results are reasonable. In addition, it can be seen from the temperature field that the area near the weld is a high-temperature area, and the central area has the highest temperature. The temperature of the contact area between the sonotrode and the copper is significantly lower. When the welding heat input is 980 J, the temperature in the center area rises sharply to 400 °C within 100 ms of the welding time. It can be seen that the energy conversion is very fast in the ultrasonic welding process. Then, the temperature slowly rises linearly with time, and the final temperature reaches 723 °C. From the Cu-ETP recrystallization temperature range combined with the simulation results, it can be seen that the temperature in the vicinity of the welding interface has reached the recrystallization temperature.

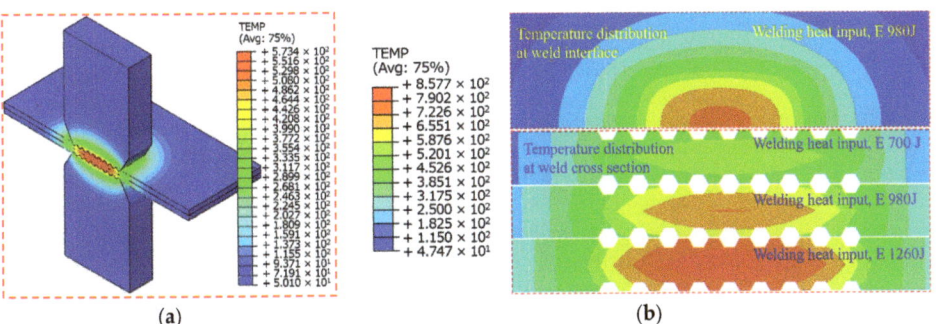

Figure 17. Temperature distribution at different welding heat inputs for the 3D plot (**a**) and xy plot (**b**).

3.4. Analysis of DRV and DRX

For the analysis of the microstructure, the texture and numerical calculation results show that the Cu-ETP welding interface produces strong shear plastic deformation and reaches the recrystallization temperature under the combined action of high frequency excitation and welding pressure during the ultrasonic welding process. The microstructure changes and connection mechanisms are dominated by DRV and DRX [27–29]. Figure 18 is a statistical graph of the EBSD recrystallization fraction of the welded joint. Blue is the recrystallized area (23.43%), yellow is the sub-structured area (14.54), and red is the deformed grain area (62.03%). It can be seen that the recrystallized areas are mostly located near the welding interface, while the subcrystals are mostly near the recrystallized grains and the deformed grains are mostly far away from the welding interface. The research results show that DRV and DRX formed a slender layered structure and ultrafine equiaxed grains, respectively, during the ultrasonic welding process.

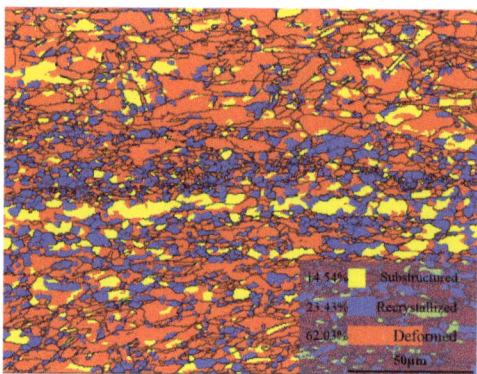

Figure 18. EBSD recrystallization fraction statistics graph of the welded joint.

Due to the combined effect of the welding pressure and ultrasonic vibration on the workpiece, the copper grains are deformed. With an increasing strain, the dislocations continue to multiply, and the dislocation density continues to increase. They gradually entangle one another or form a cellular substructure. These defects return to the undistorted state before being deformed, and the temperature rises continuously during the ultrasonic welding process, which provides an exciting energy for the recovery process. When the internal temperature of the crystal is high enough and the energy is large enough, the dislocations will migrate, the dislocation nodes will be unpinned, and the unlike dislocations will be offset when they are in contact on the new slip surface to reduce the dislocation density. It is worth noting that there is no large-angle grain boundary migration during the recovery process. DRV-driven dislocation climbing and cross-slips will form dislocation cells or subcrystals. At the same time, a few unstable crystal grains may split and form high-orientation boundaries to separate the dislocation cell masses and form elongated, layered deformation bands. Finally, the deformation bands evolved into elongated layered particles rich in LAGBs. Unstable splitting is considered to be the initial stage of ultrafine grain formation on the bonding surface. This may also contribute to the uneven distribution of fine grains at the bonding interface.

The method of dynamic recrystallization is related to the strain inside the material. When the strain is small, as shown in area I in Figure 13, discontinuous dynamic recrystallization (DDRX) mainly occurs. DDRX mainly nucleates and grows at the original grain boundary or the trigeminal node of the grain boundary. That is, the nucleation grows by arching the nucleation mode at the grain boundary. When the crystal strain is small, the dislocation density of adjacent crystal grains is different due to an uneven deformation. A grain with a large degree of deformation usually has a larger dislocation density, and a large dislocation density will increase the number of subcrystals formed inside the grain and reduce the size [18,30]. To reduce the free energy of the system, part of the subcrystals in the crystal grains with low dislocation density will migrate to the crystal grains with a high dislocation density through the adjacent grain boundary arching method and then merge the subcrystals to recrystallize the nucleus without distortion (Figure 19a). Figure 19b shows the DDRX nucleation model of grain boundary removal. The equation is:

$$\Delta E \geq \frac{2\gamma}{L} \qquad (8)$$

where γ is the surface energy of copper (approximately 1.81 J·m^{-2}); and L is the chord corresponding to the bulging arc (approximately 5 µm). This shows that the DDRX nucleation energy of copper is 7.24×10^5 J. Due to the combined effect of the welding pressure and ultrasonic vibration. The welding energy is much greater than this value, and the material will undergo plastic deformation. Therefore, ultrasonic welding provides the energy and strain conditions for DDRX nucleation.

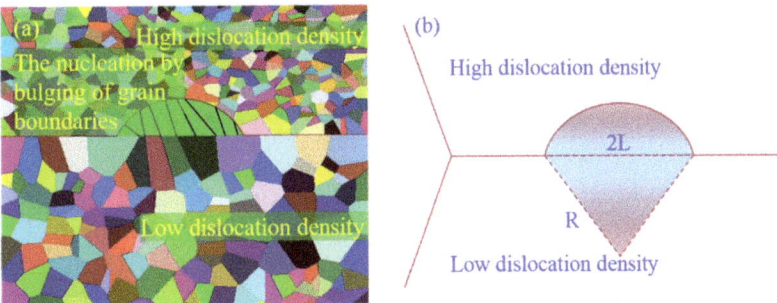

Figure 19. Schematic diagram of DDRXODF nucleation: (**a**) Nucleation by bulging of grain boundaries; (**b**) nucleation model [18,30].

Continuous dynamic recrystallization (CDRX) mainly occurs in areas with large plastic deformation (Figure 13c). Due to the high frequency friction existing in the interface area, the material in this area produces severe shear plastic deformation and obtains a higher temperature. At the same time, dislocation cells or subcrystals produced by DRV can also trigger CDRX. The process is as follows: First, the shear plastic deformation drives the sliding of dislocations within the subcrystals. To adapt to the adjacent strain incompatibility, subcrystal boundaries are combined with sliding dislocations, and the grains are refined by CDRX. Finally, the subcrystals with LAGBs become fine equiaxed CDRX grains with HAGBs through an increase in the misorientation and subgrain/grain boundary rotation. Since subcrystalline nucleation is a form of CDRX nucleation, subcrystalline grains are often accompanied by CDRX grains (Figure 18). When the metal is deformed to a large extent, the dislocations in the crystal continue to multiply, and the dislocations entangle with one another to form a cellular structure. As the temperature rises, the cell walls become flat and form subcrystals. Some of the subcrystals can merge with adjacent subcrystals to form HAGBs, which have a higher mobility and can move quickly to eliminate dislocations during movement, thereby forming a distortion-free recrystallization core. The other part of the subcrystals, due to its large misorientation, tends to migrate as the temperature rises and transforms into LAGBs before developing into recrystallization cores.

4. Conclusions

To summarize, we have demonstrated that the orthogonal test optimization method can be used to conduct ultrasonic welding tests on pure Cu with a thickness of 0.3 mm. The process parameters combination of joint welding can be evaluated and optimized by studying the joint microstructure and mechanical properties. Besides, using EBSD technology to analyze the microstructure and texture changes of the welded joints, and combining the finite element method to study the recrystallization process of copper ultrasonic welding.

(1) The micro-EDM straight groove size error is approximately 5 μm, and the surface roughness reaches 0.6 μmRa, which can form sharp corners inside the component and ensure the machining accuracy and surface quality of the micron-level 3D waveguide network.

(2) The optimal process parameters combination of Cu-ETP ultrasonic welding were obtained through experiments ($P = 1400$ W, $t = 700$ ms, $F = 1200$ N). The finite element calculation temperature peak value under this parameter reached 723 °C, which exceeded the recrystallization temperature of Cu-ETP from 379.05 to 433.2 °C. The quality of the welded joint is determined by the mechanical interlocking of the local contact surface at a low welding heat input. However, this interlocking is only a macroscopic mechanical fitting, and it is easy to detach under the action of an external force. When the energy acquired by the atom is greater than its activation energy for diffusion, as the welding heat input increases the atom diffuses, and the metal atoms interact to form a bond.

(3) There are LAGBs, subcrystalline grains and fine equiaxed grains in the Cu-ETP weld, and the average grain size is approximately 1–2 μm. These newly generated fine equiaxed grains increase the number of grain boundaries, hinder the movement of dislocations and improve the strength and plasticity of the material.

(4) The microstructure, texture and numerical calculation results show that Cu-ETP has DRV and DRX effects. The DRV and DRX effects make the weld form a slender layered structure and ultrafine equiaxed grains and play a leading role in microstructure change and adhesion mechanisms. After welding, the brass texture {112}<111> and cubic texture {001}<100> are transformed into shear textures {111}<143>, {111}<110> and {221}<122>, as well as some other textures, such as {001}<130>, {001}<230>, {114}<131>, etc. In addition there are a large number of regenerated fine cubic textures {100}<001> and shear texture {111}<110> at the welding interface.

Author Contributions: Conceived and designed the experiments, finite element simulation, performed the experiments, writing—original draft, Z.S.; funding acquisition, conceived and designed the experiments, writing—review & editing, Z.Z.; Funding acquisition, performed the experiments, Y.Z.; data analysis, writing—review & editing, H.Z.; sample fabrication, Q.X. All authors contributed to the data analysis and discussion. All authors have read and agreed to the published version of the manuscript.

Funding: This research was funded by the financial support from the National Natural Science Foundation of China (grant no. U1731118) and the Science and Technology Research Project of Jiangxi Provincial Department of Education (GJJ190918).

Institutional Review Board Statement: Not applicable.

Informed Consent Statement: Informed consent was obtained from all subjects involved in the study.

Data Availability Statement: The data presented in this study are available on request from the corresponding author.

Conflicts of Interest: The authors declare no conflict of interest.

References

1. Shan, W.; Yang, J.; Shi, S.; Yao, Q.; Zuo, Y.; Lin, Z.; Chen, S.; Zhang, X.; Duan, W.; Cao, A.; et al. Development of Superconducting Spectroscopic Array Receiver: A Multibeam 2SB SIS Receiver for Millimeter-Wave Radio Astronomy. *IEEE Trans. Terahertz Sci. Technol.* **2012**, *2*, 593–604. [CrossRef]
2. Groppi, C.; Kawamura, J.H. Coherent Detector Arrays for Terahertz Astrophysics Applications. *IEEE Trans. Terahertz Sci. Technol.* **2011**, *1*, 85–96. [CrossRef]
3. Wang, Y.; Yang, Q.; Liu, X.; Liu, Y.; Liu, B.; Misra, R.D.K.; Xu, H.; Bai, P. Microstructure and mechanical properties of amorphous strip/aluminum laminated composites fabricated by ultrasonic additive consolidation. *Mater. Sci. Eng. A* **2019**, *749*, 74–79. [CrossRef]
4. Friel, R.J.; Harris, R.A. Ultrasonic Additive Manufacturing–A Hybrid Production Process for Novel Functional Products. *Procedia CIRP* **2013**, *6*, 35–40. [CrossRef]

1. Kelly, G.S.; Just, M.S.; Advani, S.; Gillespie, J.W. Energy and bond strength development during ultrasonic consolidation. *J. Mater. Process. Technol.* **2014**, *214*, 1665–1672. [CrossRef]
2. Kelly, G.S.; Advani, S.; Gillespie, J.W.; Bogetti, T.A. A model to characterize acoustic softening during ultrasonic consolidation. *J. Mater. Process. Technol.* **2013**, *213*, 1835–1845. [CrossRef]
3. Kelly, G.S.; Advani, S.G.; Gillespie, J.W. A model to describe stick–slip transition time during ultrasonic consolidation. *Int. J. Adv. Manuf. Technol.* **2015**, *79*, 1931–1937. [CrossRef]
4. Zhang, L.; Tong, H.; Li, Y. Precision machining of micro tool electrodes in micro EDM for drilling array micro holes. *Precis. Eng.* **2015**, *39*, 100–106. [CrossRef]
5. Dehoff, R.; Babu, S. Characterization of interfacial microstructures in 3003 aluminum alloy blocks fabricated by ultrasonic additive manufacturing. *Acta Mater.* **2010**, *58*, 4305–4315. [CrossRef]
6. Sridharan, N.; Norfolk, M.; Babu, S.S. Characterization of Steel-Ta Dissimilar Metal Builds Made Using Very High Power Ultrasonic Additive Manufacturing (VHP-UAM). *Met. Mater. Trans. A* **2016**, *47*, 2517–2528. [CrossRef]
7. Gao, S.; Wu, C.; Padhy, G.; Shi, L. Evaluation of local strain distribution in ultrasonic enhanced Al 6061-T6 friction stir weld nugget by EBSD analysis. *Mater. Des.* **2016**, *99*, 135–144. [CrossRef]
8. Ward, A.A.; French, M.R.; Leonard, D.N.; Cordero, Z.C. Grain growth during ultrasonic welding of nanocrystalline alloys. *J. Mater. Process. Technol.* **2018**, *254*, 373–382. [CrossRef]
9. Lin, J.-Y.; Nambu, S.; Koseki, T. Evolution of Bonding Interface during Ultrasonic Welding between Ni and Steels with Various Microstructure. *ISIJ Int.* **2020**, *60*, 330–336. [CrossRef]
10. Siddiq, A.; El Sayed, T.; Siddiq, A. A thermomechanical crystal plasticity constitutive model for ultrasonic consolidation. *Comput. Mater. Sci.* **2012**, *51*, 241–251. [CrossRef]
11. Fujii, H.; Endo, H.; Sato, Y.; Kokawa, H. Interfacial microstructure evolution and weld formation during ultrasonic welding of Al alloy to Cu. *Mater. Charact.* **2018**, *139*, 233–240. [CrossRef]
12. Nambu, S.; Seto, K.; Lin, J.-Y.; Koseki, T. Development of a bonding interface between steel/steel and steel/Ni by ultrasonic welding. *Sci. Technol. Weld. Join.* **2018**, *23*, 687–692. [CrossRef]
13. Sriraman, M.; Babu, S.; Short, M. Bonding characteristics during very high power ultrasonic additive manufacturing of copper. *Scr. Mater.* **2010**, *62*, 560–563. [CrossRef]
14. Xie, J.; Zhu, Y.; Bian, F.; Liu, C. Dynamic recovery and recrystallization mechanisms during ultrasonic spot welding of Al-Cu-Mg alloy. *Mater. Charact.* **2017**, *132*, 145–155. [CrossRef]
15. Fujii, H.T.; Goto, Y.; Sato, Y.S.; Kokawa, H. Microstructure and lap shear strength of the weld interface in ultrasonic welding of Al alloy to stainless steel. *Scr. Mater.* **2016**, *116*, 135–138. [CrossRef]
16. Haddadi, F.; Tsivoulas, D. Grain structure, texture and mechanical property evolution of automotive aluminium sheet during high power ultrasonic welding. *Mater. Charact.* **2016**, *118*, 340–351. [CrossRef]
17. Srinivasan, V.; Balamurugan, S.; Balakarthick, B.; Darshan, S.D.; Prabhu, A.D. Experimental investigation on ultrasonic metal welding of copper sheet with copper wire using Taguchi method. *Mater. Today Proc.* **2020**. [CrossRef]
18. Elangovan, S.; Prakasan, K.; Jaiganesh, V. Optimization of ultrasonic welding parameters for copper to copper joints using design of experiments. *Int. J. Adv. Manuf. Technol.* **2010**, *51*, 163–171. [CrossRef]
19. Li, H.; Cao, B.; Liu, J.; Yang, J. Modeling of high-power ultrasonic welding of Cu/Al joint. *Int. J. Adv. Manuf. Technol.* **2018**, *97*, 833–844. [CrossRef]
20. Jedrasiak, P.; Shercliff, H.; Chen, Y.C.; Wang, L.; Prangnell, P.; Robson, J.D. Modeling of the Thermal Field in Dissimilar Alloy Ultrasonic Welding. *J. Mater. Eng. Perform.* **2014**, *24*, 799–807. [CrossRef]
21. Elangovan, S.; Semeer, S.; Prakasan, K. Temperature and stress distribution in ultrasonic metal welding—An FEA-based study. *J. Mater. Process. Technol.* **2009**, *209*, 1143–1150. [CrossRef]
22. Yang, J.; Zhang, J.; Qiao, J. Molecular Dynamics Simulations of Atomic Diffusion during the Al–Cu Ultrasonic Welding Process. *Materials* **2019**, *12*, 2306. [CrossRef]
23. Shen, N.; Samanta, A.; Ding, H.; Cai, W.W. Simulating microstructure evolution of battery tabs during ultrasonic welding. *J. Manuf. Process.* **2016**, *23*, 306–314. [CrossRef]
24. Mariani, E.; Ghassemieh, E. Microstructure evolution of 6061 O Al alloy during ultrasonic consolidation: An insight from electron backscatter diffraction. *Acta Mater.* **2010**, *58*, 2492–2503. [CrossRef]
25. Zeng, X.; Xue, P.; Wu, L.; Ni, D.; Xiao, B.; Wang, K.; Ma, Z. Microstructural evolution of aluminum alloy during friction stir welding under different tool rotation rates and cooling conditions. *J. Mater. Sci. Technol.* **2019**, *35*, 972–981. [CrossRef]
26. Fujii, H.T.; Sriraman, M.R.; Babu, S.S. Quantitative Evaluation of Bulk and Interface Microstructures in Al-3003 Alloy Builds Made by Very High Power Ultrasonic Additive Manufacturing. *Met. Mater. Trans. A* **2011**, *42*, 4045–4055. [CrossRef]

Article

The Weldability of Duplex Stainless-Steel in Structural Components to Withstand Corrosive Marine Environments

Iñigo Calderon-Uriszar-Aldaca [1,*], Estibaliz Briz [2], Harkaitz Garcia [2] and Amaia Matanza [3]

[1] TECNALIA, Basque Research and Technology Alliance (BRTA), Mikeletegi Pasealekua 2, 20009 Donostia-San Sebastián, Spain
[2] Department of Mechanical Engineering, University of the Basque Country (UPV/EHU), 48940 Leioa, Vizcaya, Spain; estibaliz.briz@ehu.eus (E.B.); arkaitz.garcia@ehu.eus (H.G.)
[3] SIAME-MPC (UPV/EHU) Université de Pau et des pays de l'Adour ISA BTP, 64600 Anglet, France; a.matanza-corro@univ-pau.fr
* Correspondence: inigo.calderon@tecnalia.com; Tel.: +34-699-907-343

Received: 12 October 2020; Accepted: 3 November 2020; Published: 5 November 2020

Abstract: There is still a considerable gap in the definition of the weldability of Duplex Stainless Steel (DSS). A lack of clarity that is explained by the standard specification of the maximum content of equivalent carbon that defines a "weldable" steel coupled with the fact that the alloying elements of DSS exceed this defined limit of weldability. In this paper, welding quality in an inert environment and in presence of chlorides is analyzed with the aim of defining optimum welding conditions of 2001, 2304, and 2205 DSS. The same procedure is followed for a hybrid weld between DSS 2205 and a low carbon mild steel, S275JR. As main output, this study defined the optimal welding conditions with tungsten inert gas without filler for each type of DSS weld that showed excellent anti-corrosion performance, with the exception of the DSS 2205-S275JR weld where widespread corrosion was observed. Additionally, this study established a relationship between the thermal input during welding and the content of alloying elements in defect-free joints. Furthermore, it demonstrated that an increase in ferrite content did not lead to a worse corrosion resistance, as expected after passivation.

Keywords: stainless steels; weldability; aggressive environments; marine environments; heat input

1. Introduction

Offshore steel structures, but also onshore in coastal areas up to a few km to the coast, suffer the effects of a harsh corrosive marine environment. This is caused mainly by moisture and chlorides that are present in the atmosphere near the sea. Chlorides cause a localized pitting corrosion attack that, in structures under tension, can cause stress concentration hot spots developing the early failure of the structure by sudden crack propagation. Thus, to face this problem, one emerging strategy is the use of Duplex Stainless Steel (DSS) instead of the simply painted carbon steel or other lesser stainless steels. DSS presents better properties, especially in terms of strength, durability, and fire resistance [1,2], when compared to the most widely used carbon structural steel in the construction industry (S275JR). These properties are due to the presence of alloying elements such as nickel and chromium, among others, generating an external protective or passive layer their microstructure, depending on the lower amount of ferrite.

Nevertheless, steel structures are composed by hot-rolled profiles and other singular elements that need to be joined to form them. When these joints are made by welding, the welding itself constitutes a localized thermal treatment that could evaporate the protective elements and change the microstructure, increasing the ferrite fraction. Thus, the protective properties of DSS could be removed by welding precisely at the more critical spots in joints, that tend to concentrate stresses. Therefore, the research on how to correctly weld these DSS with a proper thermal input is valuable to enable the safe manufacturing of steel structures under marine environment.

The main reason specific building standards, such as the Eurocodes [3–9], have yet to include the use of these steel grades is the uncertainty surrounding spots that are subjected to thermal aggressions. In other words, these standards have not yet included stainless steel in building structures, due to the complexity of establishing the parameters that can guarantee suitable and safe use of the materials after the welding process [10,11].

Consequently, research on the equivalent carbon content method specified in these standards [3–9] indicates that it is well suited to those non-alloyed steels, the so-called weldable steels in the standardization literature. Thus, the application of that method to stainless steel that has a high proportion of alloying elements is pointless, because the equivalent carbon content of them will always exceed the previously set threshold. Weldability as such, therefore, needs to be demonstrated by other alternative methods, which will ensure that the protective properties of the corresponding alloy will be locally retained within the Heat-Affected Zone (HAZ) of the weld.

Locally applied heat treatment, when incorrectly applied, could modify the welding zone of stainless steel, in such a way as to reduce the protective elements of the material and leave residual stresses. For instance, Elsaady et al. presented the behavior of different welding temperature ranges for a duplex steel [12]. The effects on the affected area made it much more vulnerable to corrosive processes such as localized pitting corrosion in a marine environment [13].

However, there is still a key clear gap between current understanding and engineering practice. According to these standards, only "weldable" steels must be used to form steel structures, even more so in marine environments, but only the equivalent carbon content method is prescribed to determine whether a steel is weldable or not, despite it discriminating DSS systematically. Thus, although DSS are in fact weldable under certain conditions and perform much better in marine environments, paradoxically they are unfairly discarded in practice because of not being "weldable" under carbon equivalent content. Therefore, there is still a need for practical and feasible alternative procedures to demonstrate the weldability of a DSS under certain conditions going beyond the equivalent carbon content method.

In fact, there are standardized procedures to see the ferrite content, or the corrosion performance under salt-spray chamber, or to do a visual examination, or to analyze a micrograph looking for intermetallic phases. Every single one of such procedures determines one single property of the DSS welding but is not able individually to determine whether a DSS is weldable or not and, even more importantly, how to do it.

Therefore, the first task is to define which properties are required to make a DSS "weldable", at least under certain conditions. The second task is to identify the more feasible way to empirically verify that such properties are meeting corresponding requirements, and define a methodology supported by standards and practice. Then, the third task is to put everything into practice, demonstrating the weldability of a single DSS by this practical way. Finally, the fourth and last task is generalization, validating proposed methodology by succeeding in applying it for more than a single DSS.

The practical significance of the work carried on in this study is that it was able to close that loop. It determined which properties are required to make a DSS weldable, and at least in equivalent terms the carbon equivalent content method does this. It proposed an alternative holistic methodology to verify such requirements. Finally, it was executed for four different types of welding with four different steels, three DSS and one carbon steel, determining the optimal thermal input and conditions for everyone.

Hence, with the aim of ensuring stainless steel properties, 1.5 mm thickness steel plates will be used for calibration tests to determine the optimum thermal input. These plates are selected as representative of cold-formed corrugated steel to which the standardized range of tests is applicable, including tests on automatic welding and welding beams with no filler metal, considered later on.

The aim here is to minimize the effects of any working variables that do not depend on the material itself avoiding introducing an additional variable relating to material compatibility. Tungsten Inert Gas (TIG) welding was chosen rather than welding, which is also widely used and might be more economical, because TIG welding without filler performs better in terms of corrosion resistance, presenting uniform welding beads and a narrower HAZ [14,15].

In addition to demonstrating the sound weldability of the flat steel plates and sheeting, these calibration tests will serve to verify the competitive properties of the plates when used alongside steel rebars, usually welded together in preassembled meshes to reinforce unique structural components. The flat steel products in the mesh often share the concrete matrix with other carbon steel rebars, placed in peripheral positions and protected only by the corresponding concrete cover. As hybrid specimens, the corrosion resistance tests of these steel products will therefore be representative of their suitability for the structural optimization of mesh reinforcements within the concrete matrices.

Accordingly, for the sake of integrating all the characteristic procedures of industrial welding, the process schedule was as follows:

1. Specimen preparation, border alignment, cleaning, bench placement, etc.
2. Welding of the specimen.
3. Brushing + pickling + passivization.

Hence, the first objective of this research is to define the optimum thermal input of each of the tested duplex stainless steels, i.e., the optimum temperature at the beginning of welding and between passes, which differs a priori for each steel grade.

Thermal input is the most important variable in any definition of properly welded stainless steel, a fact that has been confirmed in multiple studies, including those by Mohammed et al., which concluded that DSS steel 2205 tolerated a higher thermal input than austenitic steels [16], and by Asif et al., which concluded that duplex 2205 stainless steel performed better with a higher thermal input than austenitic steels [17]. Likewise, Tasalloti et al. concluded that lower heat inputs, also for DSS 2205, produced a great disparity in its composition (i.e., Cr, Ni and Mo) [18]. Besides, it is noteworthy that prolonged thermal inputs can improve the performance of welds [19]. Additionally, Asif et al. [17] underlined how lower inputs imply higher fractions of ferrite and severe precipitation of chromium nitrides; this implies a degradation of mechanical properties and corrosion resistance, such as chloride pitting corrosion, typical of marine environments.

Finally, it is worth mentioning the work carried out by Subramanian et al., who carried out an exhaustive analysis of the anode metal to cathode metal ratio and its influence on the degree of galvanic corrosion [20], to be later considered in this study for dissimilar joints. Likewise, the study by Paul et al. should be highlighted, which examined the way excessive polarization can be controlled to avoid pitting and can simultaneously reduce anode consumption [21].

Secondly, after welding with each corresponding optimum input (welding speed, temperature, etc.), the welds were validated by the full range of Non-Destructive Tests (NDT) required for the certification of welding procedures and, likewise, corrosion tests in a salt spray chamber in accordance with EN ISO 9227-NSS [22].

Although studies support the good performance of stainless steel in all types of environments [23], the equilibrium in DSS between ferrite and austenite is necessary for a good performance in marine environments, [24]. Hence, these tests successfully determined which of the DSS welded joints presented the most competitive characteristics for mechanical performance and which of them still behaved well in saline environments.

Additionally, following the same above-mentioned procedures, the feasibility of welding DSS and S275JR, a weldable type of carbon steel, was also analyzed in this research, above all with respect to galvanism. In similar studies, Sternhell et al. [25] showed the anodal role of copper in marine environments and Subramanian et al. analyzed the behavior of zinc with stainless steel in tropical marine environments [20].

Finally, as a last remark, the research value chain presents several stages. In the first stage, the novelty lies in realizing and identifying a problem that need to be solved. Then, the novelty comes when the first solutions to that problem are proposed, and when these solutions are optimized after that. Finally, at the end of the day, the novelty and the research added value still lie in performing it simply and more feasibly in a practical way. Thus, the novelty of this study is presenting an optimized and feasible procedure to determine DSS weldability in practice, alternative to the very simple carbon equivalent method. Furthermore, the novelty also lies in putting it in practice for several DSS, demonstrating simultaneously the weldability of the DSS, the relationship between their optimized thermal inputs, the feasibility of the methodology itself and its generalization for several DSS, effectively closing the loop.

2. Materials and Methods

2.1. Materials

Regarding the used materials, Table 1 shows the results of the DSS chemical composition tests of each steel grade with the aim of assessing the weldability of the different types of DSS and their corrosion resistance in marine environments. Each specimen was formed by two sheets of $180 \times 80 \times 1.5$ mm^3 dimensions, welded together by Tungsten Inert Gas (TIG) welding. Four different specimen types were built up depending on the material of each welded sheet and several samples of each type:

- Pure DSS 1.4482 (2001)
- Pure DSS 1.4362 (2304)
- Pure DSS 1.4462 (2205)
- Hybrid DSS 1.4462 (2205)/carbon steel (S275JR)

Table 1. Chemical composition of duplex stainless steels (%).

Steel	C	Si	Mn	P$_{max}$	S	Cr	Ni	Mo	N	Cu
2001 LDSS	≤0.03	≤1	4.0–6.0	35	0.03	19.5–21.5	1.5–3.5	0.1–0.6	0.05–0.25	≤1
2304 LDSS	≤0.03	≤1	≤2	35	0.015	22–24	3.5–5.5	0.1–0.6	0.05–0.25	0.15–0.6
2205 DSS	≤0.03	≤1	≤2	35	0.015	21–23	4.5–6.5	2.5–3.5	0.1–0.22	-
S275JR	21	-	1.5	0.035	35	-	0.012	-	-	0.55

2.2. Methods

The research tasks were divided into different stages, as indicated in Table 2. The first step was to determine the optimum thermal input, after this, different specimens were welded and analyzed to determine their welding quality and its behaviour in a marine environment.

Table 2. Research stages.

Stage	Item	Description
S1	Welding	Experimental characterization of the optimum thermal input for each steel grade under TIG welding conditions without filler.
S2	Welding	Performing TIG welds without filler, each material will use its optimal input.
S3	Test	Application of NDT: Visual, penetrating liquids, X-rays
S4	Test	Micrographies, microstructural analysis and ferrite content
S5	Test	Corrosion test in salt spray chamber

The base material for welding the austenitic-ferritic stainless steel was prepared in different dimensions for the post-welding tests.

Figure 1 schematizes weld join design (a) and welding sequence (b) used in each specimen. From a design point of view, a butt weld with full penetration between two 1.5 mm sheets was carried out, without edge preparation and without sheet spacing. Besides, an automatic TIG welding without filler was performed in a single pass from one side and without backing. Accordingly, this minimizes the number of variables of the procedure.

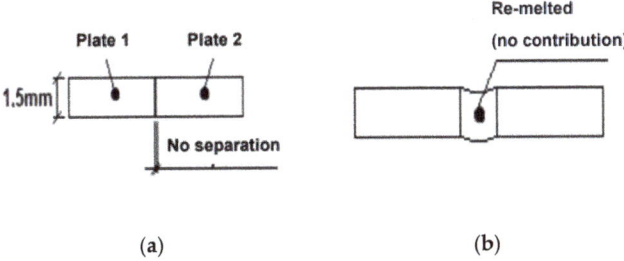

(a) (b)

Figure 1. Detail of butt weld with full penetration without edge preparation nor sheet spacing: (a) cross-section of welded joint design; (b) Welding sequences.

Table 3 summarizes the predefined variables used during the welding for the different materials. These variables are identical for the welding of all materials because they are directly associated with the welding process, regardless of the type of material to be welded.

Table 3. Values considered for predefined variables.

Variables	Value
Welding process	T.I.G. (Tungsten Inert Gas. 141)
Mechanization level	Automated
Type of welded joint	Butt weld, Single side, Backless (BW SS NB)
Welding positions	Under hand (PA)
Thickness of base sheet (mm)	1.5 mm
Outside diameter (mm)	N/A
Type of metal filler	No support
Shielding gas	99.9% Argon
Shielding gas flow rate	12 L/min
Shielding Gas Support	99.9% Argon
Support shielding gas flow rate	4 L/min
Type of electricity	Direct Current, Electrode to negative
Minimum preheating temperature	Room temperature (15 °C)
Maximum temperature between passes	N/A
Post-Welding Heat Treatment	N/A
Maximum bead oscillation/width	Straight
Pulsed welding parameters	No
Tungsten electrode, type/diameter	Tungsten + 2% Torio (Red), 2.4 mm
Arc length	3–4 mm
Gun angle	70–80°
Preparation and cleaning method	Initial cleaning: brushing with stainless steel tines Final cleaning: brushing, pickling + passivization
Welding machine	Praxair Triton 2201 AC/DC
Automation equipment	Tractor carriage BUG-O SYSTEMS
Others	Sheet clamping by tools

During stage 2, prior to the welding of the final coupons, a series of welding tests were performed to check the behavior of each base material. The purpose of these tests was to weld the coupons of each base material with different thermal inputs and to assess their specific inputs, thereby obtaining metallic welding continuity and adequate penetration through the root zone. This input should result in welds without plate perforation due to excessive input, while also avoiding the lack of fusion because of insufficient input.

Variables defined in Table 3 were the same for each specimen; however, thermal input depends on the material of specimens, so specific thermal input was defined for each specimen. Besides, it depends on current intensity, voltage, and welding speed. Thus, the following Equation (1) was used for the calculation of the thermal input, where Q is the derived thermal input in J/mm, I is the current intensity in Amperes, V is the voltage in Volts, and S_w is the welding speed in mm/s. For this study 100 A current density was defined; arc length defines voltages, considering an arc length of 3–4 mm (as Table 3 summarizes) means a voltage of 11–11.5 V.

$$Q = \frac{I \cdot V}{S_W} \tag{1}$$

The coupons were welded at different speeds, using the option provided by the equipment to regulate the speed, see Figure 2. In this way, the welding was performed with different thermal inputs and the corresponding results were compared and analyzed to define the optimum one. Initial tests showed that 2001 LDSS resists lower thermal input before drilling than 2304 LDSS and 2205 LDSS while 2304 LDSS resist a little bit more thermal input than 2001 LDSS and 2205 DSS a little bit more than 2304 LDSS. Thus, 11 tests were carried out only to optimize the thermal input of 2001 LDSS and, taking this as a reference, the thermal input was slightly increased step by step for 2304 LDSS and 2205 LDSS thereafter, until reaching their own optimum input.

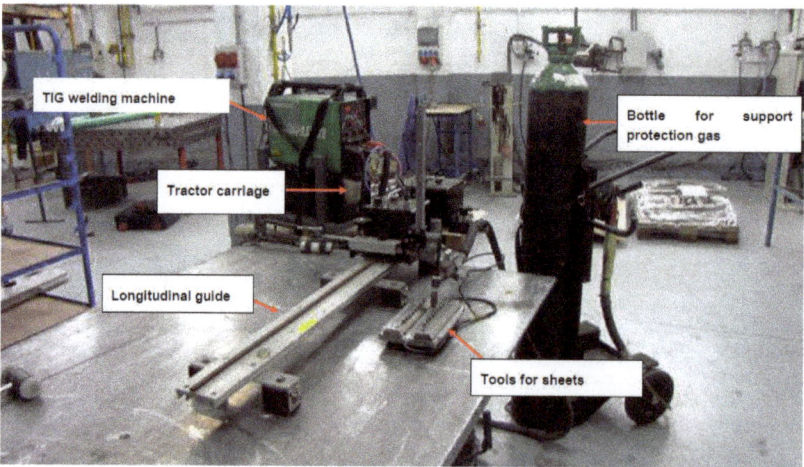

Figure 2. Automatic welding equipment: including sheet tooling, traction carriage on longitudinal guide rail to automatically control welding speed, Tungsten Inert Gas (TIG) welding machine to apply constant thermal input parameters, and shielding gas connection to perform welding under inert gas.

Thermal input also can be defined from nickel equivalent and chromium equivalent numbers. For instance, using the Schaeffler equation [26], a direct relationship can be derived between the summation of equivalent nickel and equivalent chromium of each DSS steel grade and the corresponding optimal thermal input of its welding seams. See Equation (2).

Then, once each coupon was welded at different welding speeds, the weld seam and adjacent area was pickled and passivated, to restore the corrosion resistance properties of the steel after welding and grinding. A pickling and passivation paste was used for this purpose. Figure 3 shows the coupons following this treatment process. Finally, following the application of these restorative treatments, the saline spray tests were performed to ensure post-welding corrosion resistance.

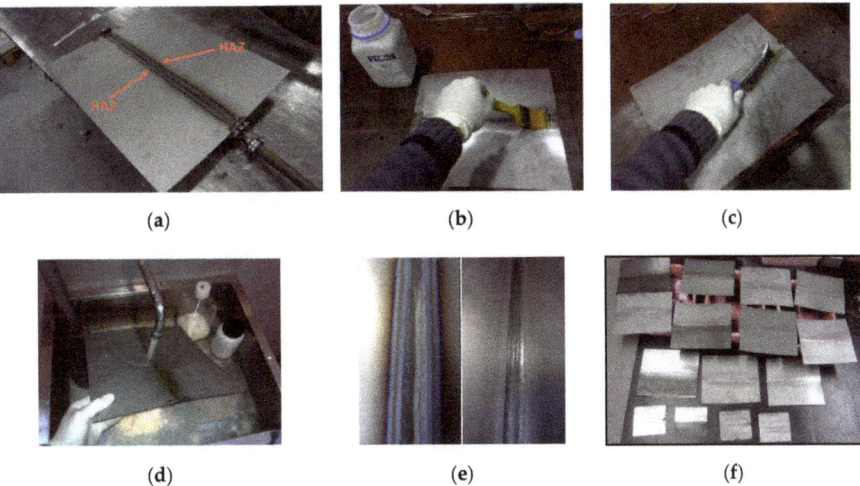

Figure 3. Welded coupons for testing, already prepared after pickling and passivation: (**a**) Detail of the HAZ appearing after welding; (**b**) Application of pickling and passivation product; (**c**) Removing the pickling and passivation product; (**d**) Sample and weld seam cleaning; (**e**) Comparison of the welded seam before and after pickling and passivation; (**f**) Welded coupons prepared for testing.

Every coupon was then analyzed by non-destructive tests (NDT) to determine the optimal thermal input for each steel as the minimum thermal input required for welding without any fault, according to the following standards.

- Visual test: as per standard ISO-17637:2011 [27], in replacement of standard EN-970:1997 [28].
- Test for the detection of surface cracks by penetrating liquids: as per standard EN-571-1:1997 [29].
- Radiographic test: as per standard EN-1435:1998 [30].

Besides, standard EN 5817 [31] was used for the result assessment; the standard defines three different quality level (D,C,B) on the basis of type, size and amount of selected imperfection. B means the highest quality and D the lowest. Accordingly, only B level was accepted in this study

The tests to derive ferrite content is based on ASTM E562 standard [32]. In order to evaluate the welded joints, the following specimen preparation and test application tasks are applied:

1. Saw cutting, in a mechanical workshop, of welding section.
2. Specific preparation of specimens, at the laboratory level, by embossing the samples in resin and mirror polishing.
3. Electrolytic attack of samples using 40% NaOH soda, to calculate the ferrite content on micrographies,
4. Electrolytic attack of samples by 10% oxalic acid, for microstructural observation in light microscope.

These tests are followed by destructive neutral salt spray (NSS) according to ISO 9227 [22]. Therefore, the samples were sprayed in a 5% sodium chloride solution within the pH range of 6.5 to 7.2 in a controlled environment, with an exposure duration up to 216 h. Figure 4 shows the coupons placed inside test chamber

Figure 4. Samples of Tungsten Inert Gas (TIG) welded stainless-steel sheets (2001, 2304, 2205 and S275-2205) during the application of the marine corrosion test in a salt spray chamber.

3. Results

This section discloses the results of the NDT for determination of optimized thermal input of each steel joint type, the results regarding the ferrite content by means of micrographies and microstructural analysis and the results of the salt spray chamber destructive tests.

3.1. Optimized Thermal Input

The process for the determination of thermal input of DSS 2001 is analyzed below. The process followed for the other materials was the same.

As mentioned before, the voltage derived from the applied arc height ranged from 11–11.5 V. A mean voltage of 11.25 V was taken into account for the calculation of the thermal input by Equation (1). A total of 11 DSS coupons were welded. Table 4 summarizes the thermal input value used for each specimen relating to 1.4482 (2001) DSS coupons.

Table 4. Therma input used for each coupon of 1.4482 (2001).

Ref. Coupon	Q (kJ/mm)
2001 × 180-A	0.256
2001 × 180-B	0.188
2001 × 180-C	0.150
2001 × 180-D	0.138
2001 × 180-E	0.321
2001 × 180-F	0.275
2001 × 180-G	0.307
2001 × 180-H	0.181
2001 × 180-I	0.231
2001 × 180-J	0.245
2001 × 180-K	0.281

For 2001 × 180-D and 2001 × 180-E coupons, the penetrating liquid and radiography tests were not applied, since they presented lack of root penetration and perforation, respectively, during visual examination. A and G coupons failed to pass the evaluation of the radiographic test, due to detected internal fusion faults. The remaining coupons met the requirements of the radiographic test. However, visual observation of the weld-seam surface should also be considered. See Figure 5 showing X-ray and penetrating liquids NDT procedure for some samples.

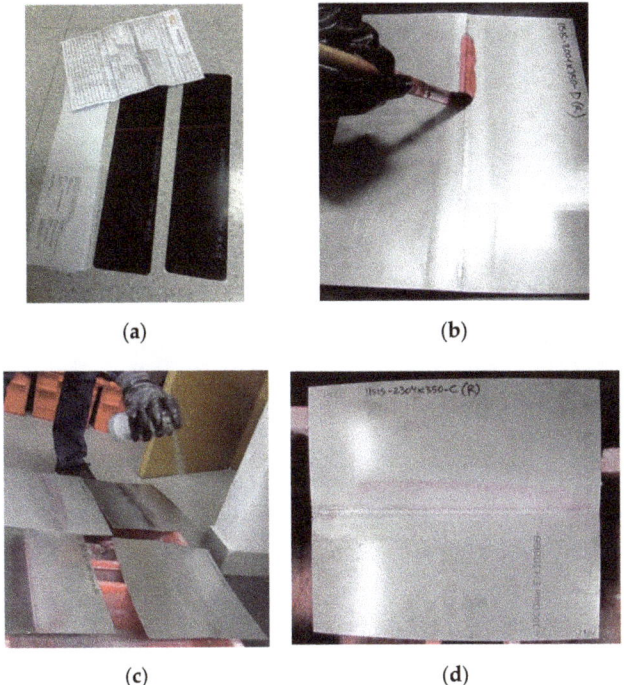

Figure 5. X-ray and penetrating liquids non-destructive testing performed on samples: (**a**) X-ray plates and test certificates of samples; (**b**) Application of penetrating liquids in red; (**c**) Application of revealing product after cleaning; (**d**) 2304 Duplex Stainless Steel (DSS) sample showing a good result with no penetration.

Following the completion of the radiographic test, a visual examination of the coupon welding surfaces was completed (Figure 6 shows the status of some welded specimens). The weld seam surface of coupon "2001 × 180-K" was unstable and irregular, with a sinusoidal weld along the entire joint. 2001 × 180-F coupon ref. showed a longitudinal contraction towards the inside of the seam on the welding side; this contraction appeared to be a kind of root concavity defect. Coupon "2001 × 180-J" was welded with a slightly lower thermal input than the coupon "2001 × 180-A". The surface showed an irregular and sinusoidal weld. It was therefore discarded as a non-optimal weld. The seam on coupon ref. "2001 × 180-B" showed an irregularly shaped bead at the welding face side, so the solidification of the seam is not optimal and could be improved. On the other hand, coupons with ref. "2001 × 180-H" and "2001 × 180-C" showed an excessively narrow weld seam width in the root area. This effect is attributed to the application of an excessively high welding speed on both coupons, and therefore, an excessively low thermal input. Table 5 summarizes radiographic results.

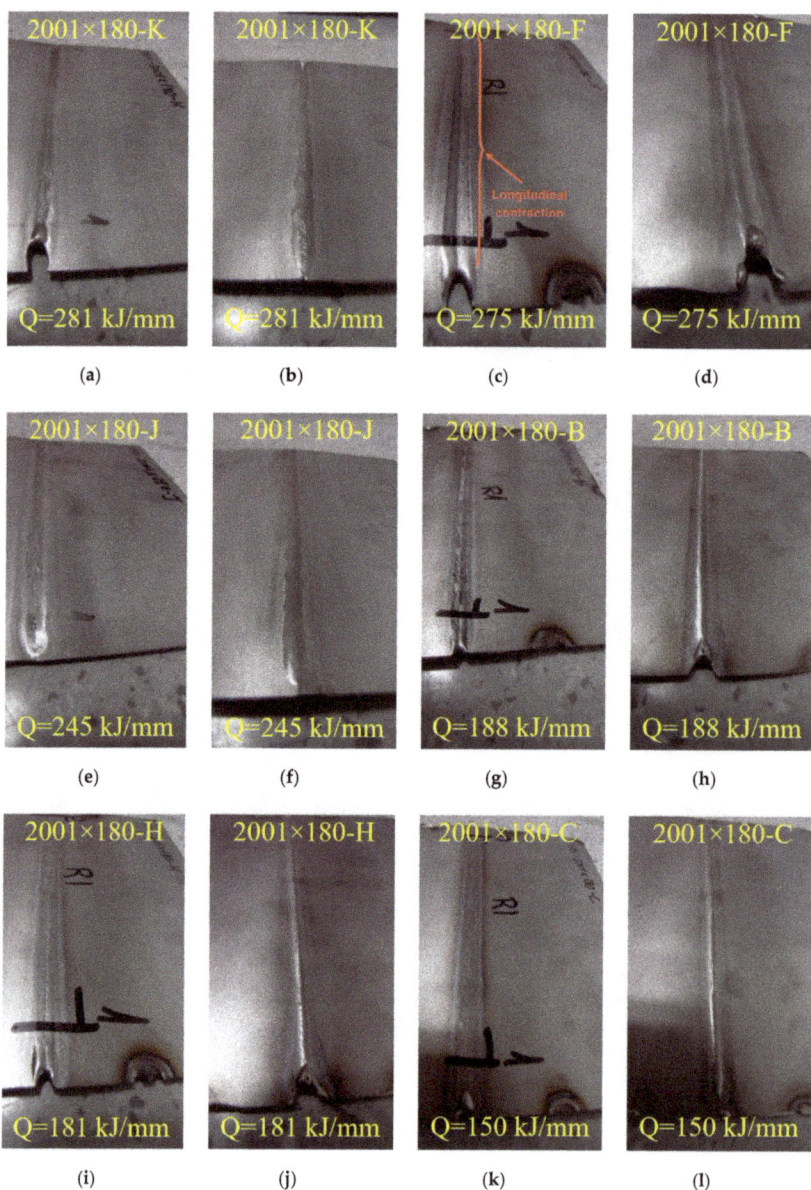

Figure 6. Welded coupons of 2001 steel from welding face and root sides: (**a**) 2001 × 180-K from the welding face side (Q = 281 kJ/mm); (**b**) 2001 × 180-K from the welding face side (Q = 281 kJ/mm); (**c**) 2001 × 180-F from the welding face side (Q = 275 kJ/mm); (**d**) 2001 × 180-F from the welding face side (Q = 275 kJ/mm); (**e**) 2001 × 180-J from the welding face side (Q = 245 kJ/mm); (**f**) 2001 × 180-J from the welding face side (Q = 245 kJ/mm); (**g**) 2001 × 180-B from the welding face side (Q = 188 kJ/mm); (**h**) 2001 × 180-B from the welding face side (Q = 188 kJ/mm); (**i**) 2001 × 180-H from the welding face side (Q = 181 kJ/mm); (**j**) 2001 × 180-H from the welding face side (Q = 181 kJ/mm); (**k**) 2001 × 180-C from the welding face side (Q = 150 kJ/mm); (**l**) 2001 × 180-C from the welding face side (Q = 150 kJ/mm).

Table 5. Radiographic results and observations depending on the thermal input.

Ref. Coupon	Q [kJ/mm]	Radiography Test	Remarks
2001 × 180-E	321	Not applicable	Drilling
2001 × 180-G	307	Not acceptable	X-ray: Lack of internal fusion
2001 × 180-K	281	OK	Irregular weld seam
2001 × 180-F	275	OK	Longitudinal contraction per side
2001 × 180-A	256	Not acceptable	X-ray: Lack of internal fusion
2001 × 180-J	245	OK	Irregular weld seam
2001 × 180-I	231	OK	Weld seam without relevant indications
2001 × 180-B	188	OK	Weld bead that can be improved per face area
2001 × 180-H	181	OK	Excessively narrow weld bead per root
2001 × 180-C	150	OK	Excessively narrow weld bead per root
2001 × 180-D	138	Not applicable	Lack of root penetration

According to the results from the visual inspection, liquid penetration test and radiographic test, I coupon was defined as the optimum weld with a thermal input of 0.231 kJ/mm. Table 6 summarizes the optimal thermal input obtained by the same procedure for different materials and parameters used in weld process.

Table 6. Summary of welding parameters and corresponding optimized thermal input.

MATERIAL	I (A)	V_{mean} (V)	Forward Speed (mm/s)	Q (kJ/mm)
2001	100	11.5	4.86	0.231
2304	100	11	3.76	0.293
2205	100	11.75	3.67	0.320
2205-S275	100	12	3.61	0.332

3.2. Ferrite Content

Prior to the microstructural analysis, a macrograph of each type of material is taken, in which base materials are observed, with corresponding heat affected zones (HAZ) and the weld.

Subsequently, a microstructural analysis is carried out including images of the cord, the HAZ and the base material. At least six micrographs are obtained for each type of material at ×400 magnification. The ferrite content and the possible presence of intergranular precipitates are analyzed. Thus, two types of micrographs are obtained:

- Micrographs of samples electrolytically etched with 40% NaOH soda: for the calculation of the ferrite content. These micrographs are displayed in color. Austenite is visualized in white, and ferrite in color.
- Micrographs of samples electrolytically attacked with 10% oxalic acid: for microstructural observation in an optical microscope (possible precipitates, intermetallic phases, etc.). These micrographs are displayed in black and white.

3.2.1. Macrographies

Figure 7 discloses the macrographies of the cross section corresponding to the four steel joint types belonging to 2001, 2304, 2205 and 275-2205 base materials. Each shows a correct welding zone with minimum drop to the root and clearly showing different aspects for the welding, base material and HAZ transition zones that indicate the presence of different phases and material inclusions. Therefore, since the welding can act as a thermal treatment evaporating the protective elements of the stainless steels disclosed at Table 1 and changing its microstructure, a micrography and microstructural analysis is required to ensure the welding process is not depriving the corrosion resistance of such materials to marine environments full of chlorides.

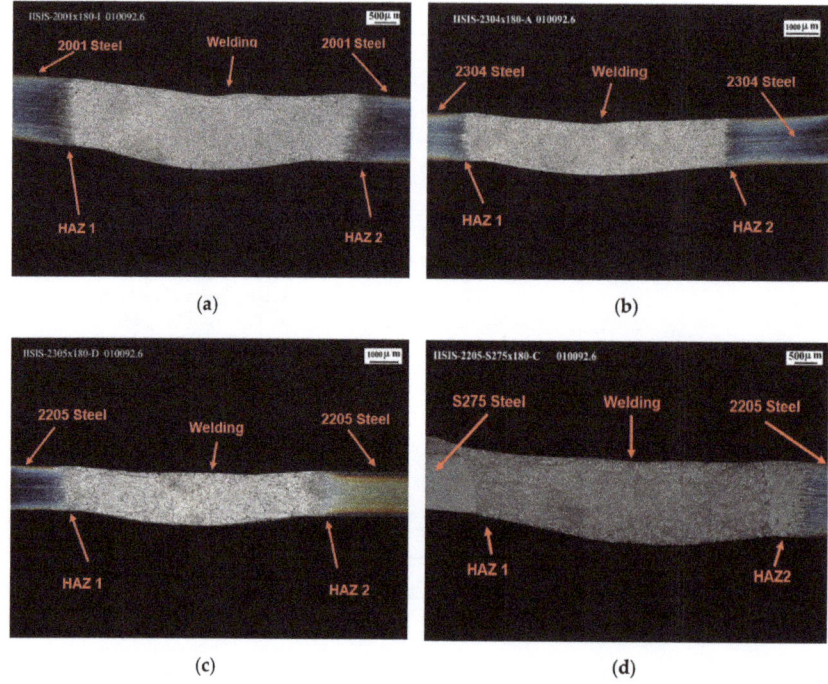

Figure 7. Cross-Section welded joint macrographies: (**a**) 2001-2001 steel joint with ×12 Magnification; (**b**) 2304-2304 steel joint with ×9 Magnification; (**c**) 2205-2205 steel joint with ×8 Magnification; (**d**) 275-2205 steel joint with ×9 magnification.

3.2.2. Micrographies

As already mentioned, the micrographies have been performed on previously etched samples with 40% NaOH soda. These micrographs are displayed in color, with austenite shown in white, and ferrite is colored. Accordingly, Table 7 summarizes the ferrite content obtained for each welded coupon after applying the test, depending on the zone.

Table 7. Ferrite content of each optimized coupon (%).

Reference	Base Steel	HAZ	Welding
2001 × 180-I	53.0	68.4	73.5
2304 × 180-A	55.6	71.6	82.4
2205 × 180-D	59.8	74	86.3
2205-S275 × 180-C	60.2	72.5	47.5

Besides, Figures 8–11 show the micrographies with ×400 magnification of the four steel joint types considered in this study. As general remarks for all cases, ferrite and austenite bands can be seen in the base material, and are elongated in the direction of the sheet hot-rolling. In welding zones, an increase in the percentage of ferrite is appreciated when compared to the base material, and it can be observed the presence of acicular austenite too. The same is applicable for the HAZ, in which the grains present a greater orientation to the hot rolling direction at the area closest to the base material (left).

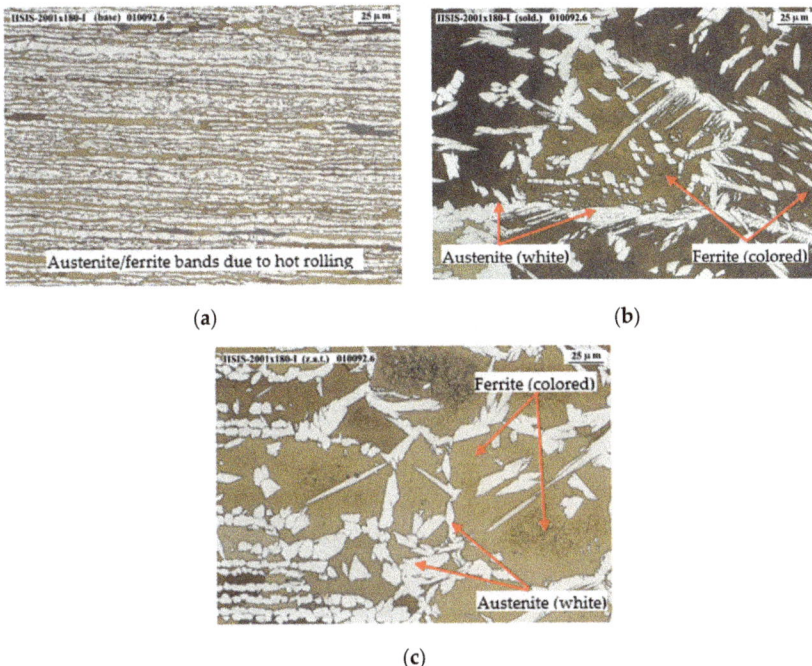

Figure 8. 2001-2001 steel joint micrographies: (**a**) Base 2001 steel with ×400 magnification; (**b**) Welding zone with ×400 Magnification; (**c**) 2205 HAZ zone with ×400 magnification.

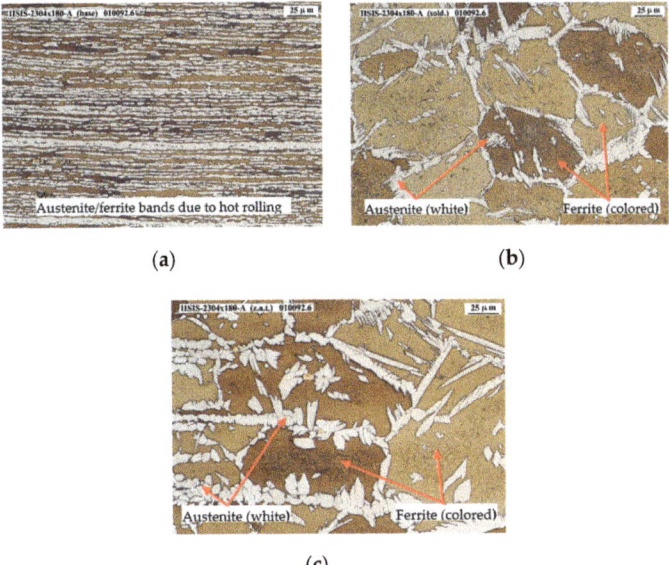

Figure 9. 2304-2304 steel joint micrographies: (**a**) Base 2304 steel with ×400 magnification; (**b**) Welding zone with ×400 Magnification; (**c**) HAZ zone with ×400 magnification.

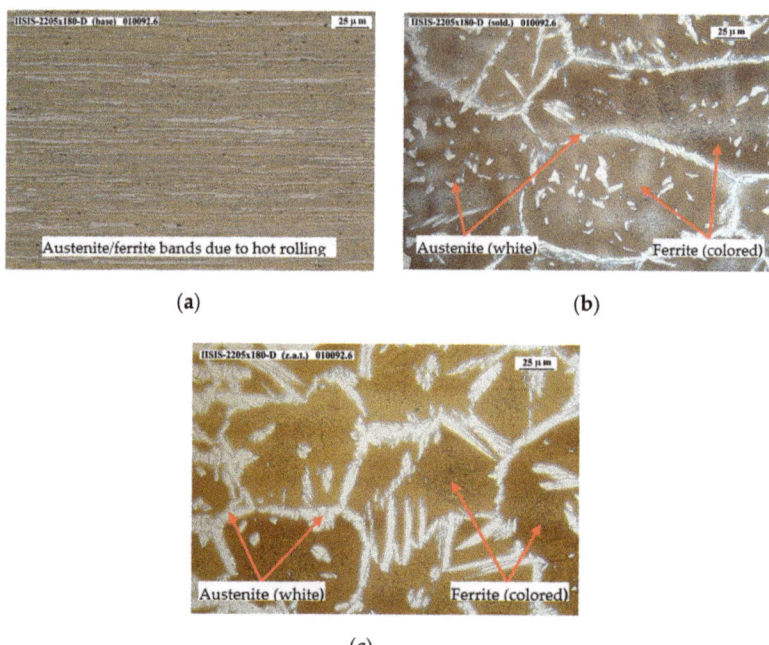

Figure 10. 2205-2205 steel joint micrographies: (**a**) Base 2205 steel with ×400 magnification; (**b**) Welding zone with ×400 Magnification; (**c**) HAZ zone with ×400 magnification.

Additionally, for the hybrid 275-2205 steel joint, see Figure 11, the ferrite appears in orange in the area of 2205 base material and corresponding HAZ. Nevertheless, the ferrite appears bluish in welding area. This difference in color is due to the fact that in the welding area, as no filler material has been used, there has been a dilution between the austeno-ferritic material 2205 and the carbon steel S275, and when this area is attacked with soda, the ferrite has reacted turning blue.

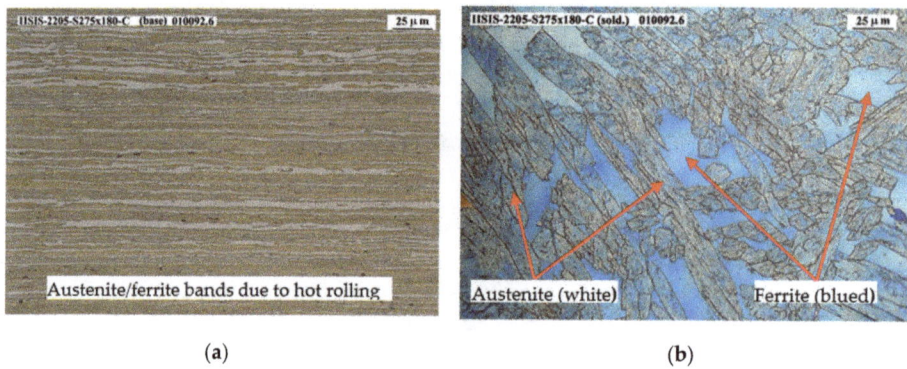

Figure 11. *Cont.*

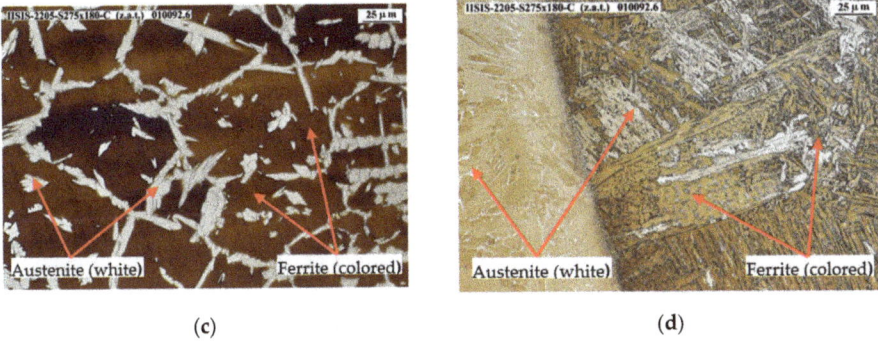

(c)   (d)

Figure 11. 275-2205 steel joint micrographies: (**a**) Base 2205 steel with ×400 magnification; (**b**) Welding zone with ×400 Magnification; (**c**) 2205 HAZ zone with ×400 magnification; (**d**) S275 HAZ zone with ×400 magnification.

3.2.3. Microstructural Analysis

The next micrographs have been performed on samples electrolytically attacked with 10% oxalic acid to enable the microstructural observation in an optical microscope (possible precipitates, intermetallic phases, etc.). Accordingly, these micrographs are displayed in black and white, see Figures 12–15, where ferrite shows a dark color and austenite a lighter one.

As an additional clarification, this kind of micrograph is used to look for possible indications of the presence of precipitates, intermetallic phases, sigma or chi phases and the like. These indications take the form of discontinuities with surrounding metallic structures. Nevertheless, despite it being a good way to see if a certain sample shows enough of such indications, it is not suitable to determine the nature of the discontinuity itself. For that purpose, the scanning electron microscope (SEM) is used as a complementary tool to complete the analysis instead. Thus, typically the more cost-effective strategy is to first analyze the samples already attacked with 10% oxalic acid in search of indications to determine the usefulness of a further SEM analysis or lack thereof, discarding it or not accordingly.

However, the micrographs of this study have shown a rather low and scarce amount of such indications, meaning almost no presence of possible precipitates or intermetallic phases. The red circles in some of the next figures just show places of special interest because of the possible presence of such indications. Nevertheless, after a further study of such places, it was considered that micrographs were not presenting enough evidence to make a SEM study advisable, discarding it accordingly.

Figure 12a–c shows the micrographies corresponding to 2001 base material, welding and HAZ zones, respectively. In these last two zones, precipitates or intermetallic phases are appreciated in some areas of the grain edges, in small quantities, such as the sigma phase or the chi phase. It should be mentioned that the appearance of the sigma phase cannot be confirmed by optical microscope means only, but rather a SEM would have to be used. However, the number of indications detected is considered low, so that even with SEM electron microscope analysis it would be difficult to define the nature of these indications.

For 2304 steel, a slightly higher quantity of this type of indications can be seen than in the 2001 material, but is equally scarce, so that even with the analysis by SEM electron microscope it would be difficult to analyze the nature of said indications. See Figure 13.

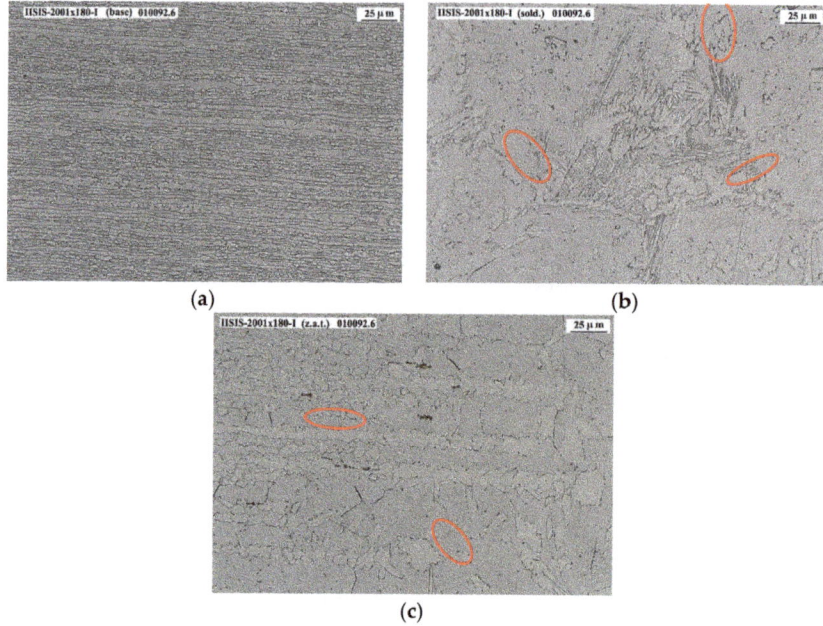

Figure 12. 2001-2001 steel joint micrographies: (**a**) Base 2001 steel with ×400 magnification; (**b**) Welding zone with ×400 Magnification; (**c**) HAZ zone with ×400 magnification.

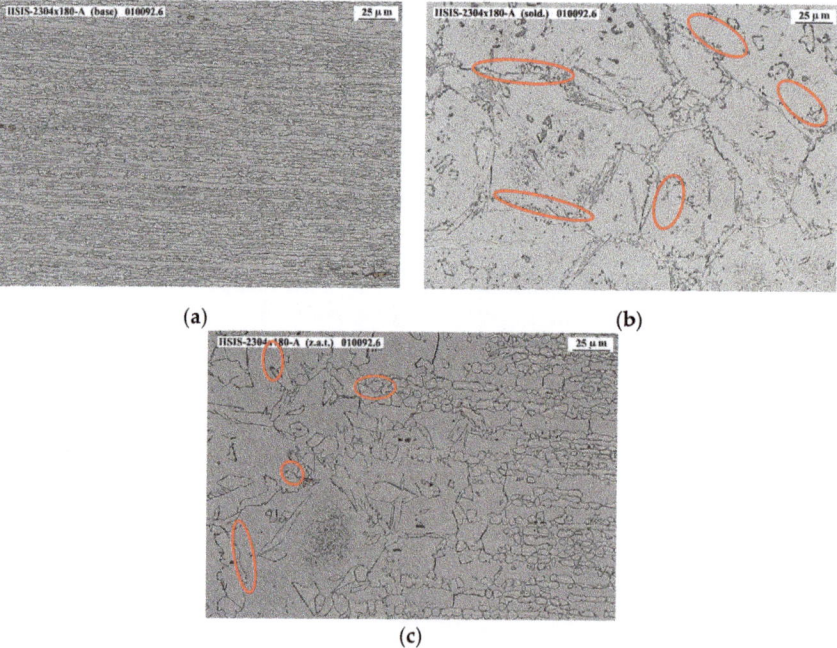

Figure 13. 2304-2304 steel joint micrographies: (**a**) Base 2304 steel with ×400 magnification; (**b**) Welding zone with ×400 Magnification; (**c**) HAZ zone with ×400 magnification.

For 2205 steel, a lower amount of this type of indications is seen than in materials 2001 and 2304, see Figure 14.

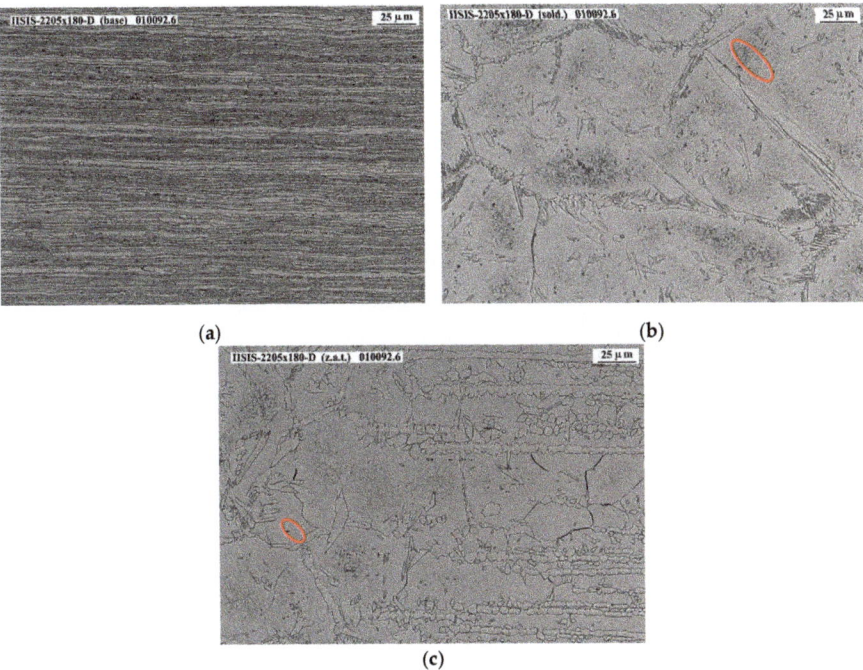

Figure 14. 2205-2205 steel joint micrographies: (**a**) Base 2205 steel with ×400 magnification; (**b**) Welding zone with ×400 Magnification; (**c**) HAZ zone with ×400 magnification.

Finally, a lower amount of this type of indications is appreciated in S275-2205 joint than in the materials 2001 and 2304 and a similar amount to that appreciated in the coupon of material 2205. See Figure 15.

Figure 15. *Cont.*

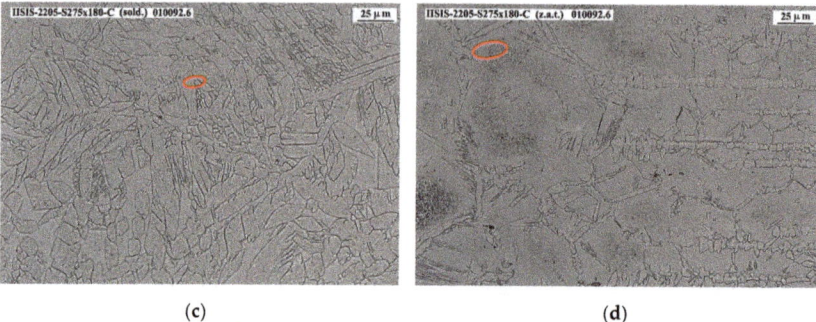

(c)   (d)

Figure 15. 275-2205 steel joint micrographies: (**a**) Base 2205 steel with ×400 magnification; (**a**) Base 275 steel with ×400 magnification; (**c**) Welding zone with ×400 Magnification; (**d**) HAZ zone with ×400 magnification.

3.3. Salt Spray Chamber Test

During this test, the samples were exposed to a harsh environment in a climatic chamber. Thus, after 216 h of exposure to an environment with 5% sodium chloride at a pH ranging from 6.5 to 7.2, it was observed that the welded samples of DSS materials under refs. '2304 × 180-A', '2205 × 180-D' and '2001 × 180-I' remained in perfect state, unaffected by corrosion. See Figure 16.

Conversely, the sample weld between DSS 2205 and the carbon steel, S275, under ref. '2205-S275 × 180-C' showed a continuous progression of red chloride corrosion in the S275 carbon steel area. See Figure 17. According to salt-spray chamber standardized testing procedure, the corrosion stages need to be checked after 24, 48, 168 and 216 h, and the results can be summarized in next steps:

- 24 h: Red corrosion on the right side of the weld. Additionally, a red corrosion spot on the left surface.
- 48 h: No significant variations, progression of red corrosion front to the left surface.
- 168 h: No significant variations, progression of red corrosion front to the left surface continues growing.
- 216 h: Spalling and mass loss due to progression of red corrosion.

Therefore, according to the results, the welding of DSS appears suitable for structural elements placed under marine environments when such welding is done with the proper thermal input. However, the welding of such DSS with carbon steel is not advisable, as it turns the carbon steel into a sacrificial anode. In view of such a result, those steels can be considered as weldable DSS steels, even though they surpass the equivalent carbon maximum content defined in current standards.

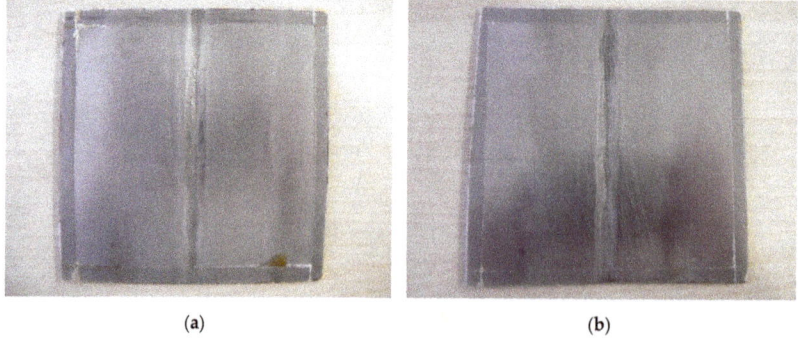

(a)   (b)

Figure 16. *Cont.*

(c)

Figure 16. DSS welded samples under simulated marine environment in salt-spray chamber test after 216 h of exposition: (**a**) 2304 DSS showing no significative variations; (**b**) 2205 DSS showing no significative variations; (**c**) 2001 DSS showing no significative variations.

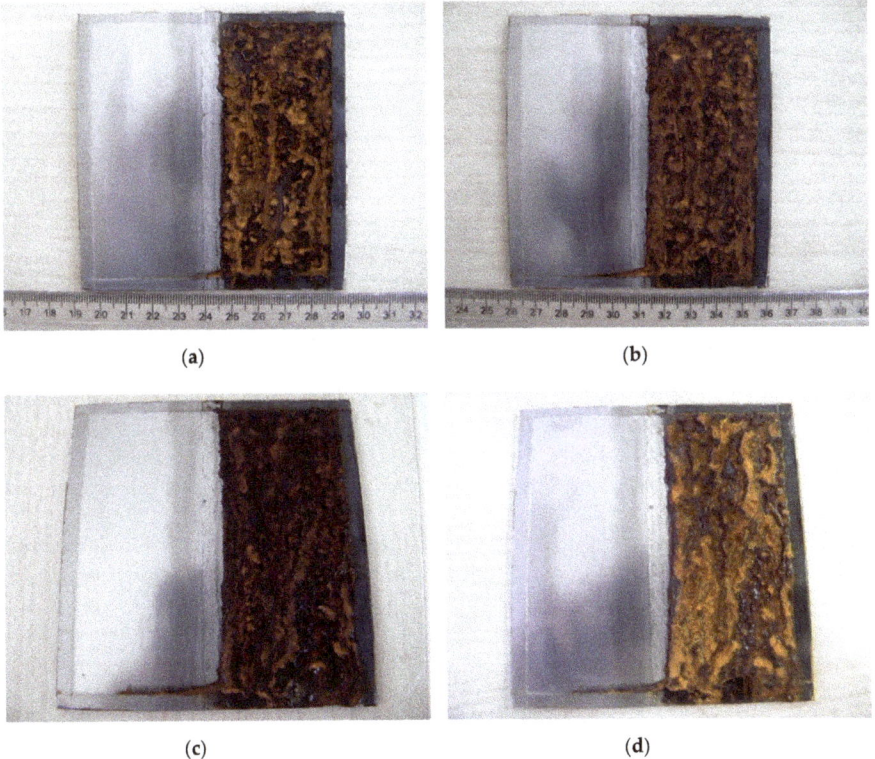

Figure 17. Hybrid 2205-S275 welded sample under simulated marine environment in salt-spray chamber test: (**a**) After 24 h of exposition showing corrosion at S275 carbon steel; (**b**) After 48 h of exposition showing corrosion progression; (**c**) After 168 h of exposition; (**d**) After 216 h of exposition showing corrosion at S275 carbon steel with aggravated mass loss and small progression towards 2205 steel.

4. Discussion

First, in view of the results disclosed in previous section, every steel joint type was able to be executed at a certain optimized thermal input that was found during the tests. This optimized thermal input is the minimum input required to ensure a suitable and faultless welding seam according to NDT (visual examination, penetrating liquids and radiographic means). Nevertheless, in case of DSS needing to resist chloride corrosion when exposed to a marine environment, there is a risk that such welding processes could cause a protective element loss and ferrite formation weakening the welding and heat affected zones. The higher the thermal input, the higher the risk, so a minimum thermal input while still enough to ensure correct welding is a key founding, but it also requires additional verification of the ferrite formation and remaining corrosion resistance after welding.

Hence, macrographies and micrographies for microstructural analysis have been performed and the results show clearly separated zones corresponding to base metal, HAZ and welding zone, see Figure 7. Ferrite content in base steel was found to be around 53–60%, with 68–72% at HAZ and 73–86% at welding zone. However, an unexpected decrease in the percentage of ferrite is observed in the welding zone of S275-2205 hybrid steel joint in comparison with the austenitic-ferritic base material 2205, up to 47.5%, due to the dilution between both base materials because of the welding without filler, that turned that micrograph to blue instead, see Figure 11b.

Nevertheless, despite the unavoidable rise in ferrite percentage caused by the welding process, although mitigated by optimized thermal input, every DSS still counts on the protective elements in their chemical composition, see Table 1, that protect it from corrosion. Thus, the true performance of each weld needs to be tested in corrosion scenario to see how well each welded joint behaves to corrosion. Accordingly, salt-spray chamber tests have been performed on each steel joint, finding that every steel joint conserve their corrosion resistance with the only exception of the hybrid steel joint S275-2205. In the case of this last joint, the early corrosion attack at the S275 steel half up to the welding seam showed a possible sacrificial anode behavior and the key aspect of the remaining protective elements of DSS after welding, since despite the much lower ferrite content, it showed accelerated corrosion.

Second, adapting the Schaeffler formula [26] with the data gathered in Table 8, obtained directly from DSS chemical composition already disclosed in Table 1, shows a well fitted relationship between the summation of equivalent nickel Ni_{eq} and equivalent chromium Cr_{eq} of each DSS steel grade with the corresponding optimal thermal input of its welding seams. It shows a good correlation under the initial boundary conditions, i.e., butt welding of two narrow 1.5 mm thick steel sheets with full penetration and neither spacing nor edge preparation. Thus, Figure 18 shows the linear regression performed to derive the terms of Equation (2).

Following this procedure, the optimal thermal input Q in KJ/mm of any DSS can therefore be derived from Equation (2), simply by introducing the summation of equivalent Chromium and Nickel $Cr_{eq} + Ni_{eq}$, in accordance with the Schaeffler formula.

$$Q = 0.0102 \cdot (Ni_{eq} + Cr_{eq}) - 0.1144 \qquad (2)$$

Third, the absence of macrodefects does not necessarily mean that the resulting weld strength is good enough. For that purpose, additional strength tests are required to ensure the welding is performing at least as good as the base materials. Besides, cyclic loadings can cause the appearance of such macrodefects, even later unleashing corrosion fatigue problems [33,34], and even permanent tensions can cause development of stress corrosion cracking if weld bead suddenly changes geometry or accelerates corrosion fatigue deterioration [35,36].

Nevertheless, regardless of such final strength, this study is still enough to demonstrate that a proper weld design, considering this strength, should be suitable for the whole service life, since the applied thermal input does not imply protective properties loss in the weld and HAZ zones.

Finally, as a last comment, following an analysis of the results on optimized thermal input obtained from the tests of the different coupons made of the three DSS in this study, it was concluded that DSS 2001 could not be welded with inputs as high as those for 2304 and 2205 steels, because the DSS 2001 material began to suffer perforations more easily than the other two as the input increased. Besides, evidence showed that 2205 DSS was also able to tolerate a slightly higher input than 2304 with no perforation. See Table 6.

Table 8. Thermal input to nickel and chromium equivalent content correlation.

Steel Grade	Ni_{eq}	Cr_{eq}	$Ni_{eq} + Cr_{eq}$	Q (KJ/mm)
2001	12	22,145	34,145	231
2304	134	261	395	293
2205	15	28	43	32

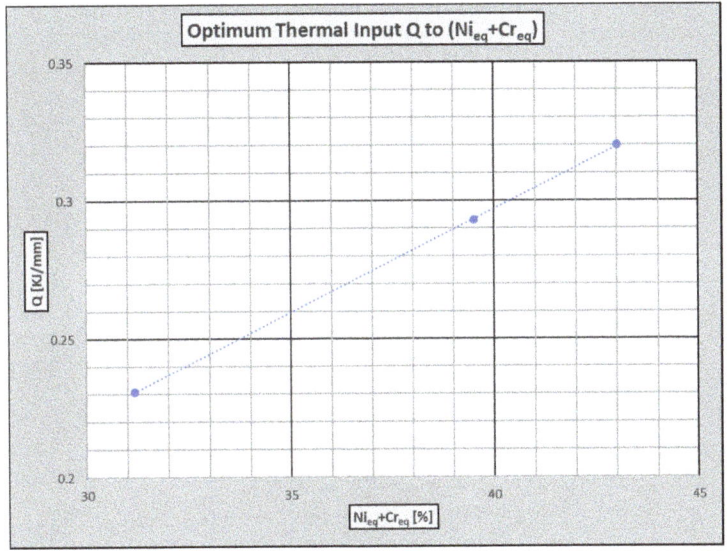

Figure 18. Optimum thermal input depending on equivalent Nickel and Chromium content.

5. Conclusions

1. Four types of Duplex Stainless Steel (DSS) joint samples, namely 2001 to 2001, 2304 to 2304, 2205 to 2205, and 2205 have been welded to carbon steel S275 for testing purposes. These DSS samples exceeded the maximum equivalent carbon rate for their theoretical classification as weldable steels in accordance with current structural standards [3–9], so an alternative procedure has been developed to ensure weldability.
2. Each DSS coupon sample has been welded holding certain boundary conditions constant, such as TIG welding without filler, no backing, no spacing, and no edge preparation, with automatic advance, and the same current intensity and voltage. Hence, the thermal input variation was simply controlled by varying only the welding speed. Several coupons of each material were therefore prepared with different thermal inputs.
3. Every coupon of the same DSS has been studied by means of non-destructive testing (NDT), according to the standard specifications for the certification of a welding procedure. Namely, visual inspection, radiographic X-ray tests and penetrating liquid tests. Those NDT were then

used to define the lowest optimum thermal input for welding an acceptable seam. Table 4 summarizes the optimum thermal input of each material.

4. Finally, to ensure remaining corrosion resistance to marine environments and chloride corrosion after welding, each specimen has been studied by means of a macrography to separate the base material, welding and heat-affected zones. Then several micrographies to obtain the ferrite content at each zone and detect indications of sigma, chi or other elements during a microstructural analysis, and finally a destructive accelerated corrosion test in a salt-spray chamber.

5. This study established a relationship between the heat generation during welding and the content of alloying elements in defect-free joints. Furthermore, it found that an increase in ferrite content did not lead to a worse corrosion resistance, as expected after passivation.

6. A salt spray test within a climatic chamber was performed on the different types of welded coupons according to EN ISO 9227 [22], using the Neutral Salt Spray (NSS) test, lasting up to 216 h, in order to demonstrate that the optimum thermal inputs had no effect on the corrosion resistance properties within the heat affected and welding zones. Following the application of the test, it was observed that the welded samples of DSS material 2001, 2304 and 2205 were corrosion free. In contrast, a continuous progression of red chloride corrosion was observed in the hybrid sample of combined materials 2205-S275, on the carbon steel half of the S275 coupon.

7. The results have shown suitable welding seams and an adequate weldability of the DSS materials 2001, 2304, and 2205 under the previously specified welding conditions, in terms of metallic continuity and corrosion behavior.

8. The hybrid welded joint of DSS steel 2205 and carbon steel S275 showed a bad performance under chloride environment as the carbon steel underwent appreciable corrosion, presumably when the joint became a galvanic couple, acting as a sacrificial anode.

Author Contributions: Conceptualization, I.C.-U.-A.; methodology, I.C.-U.-A.; formal analysis, E.B. and H.G.; investigation, A.M.; resources, I.C.-U.-A.; data curation, H.G. and E.B.; writing—original draft preparation, H.G. and E.B.; writing—review and editing, I.C.-U.-A.; visualization, A.M.; supervision, I.C.-U.-A. and A.M.; project administration, I.C.-U.-A.; funding acquisition, I.C.-U.-A. All authors have read and agreed to the published version of the manuscript.

Funding: This research was funded by ACERINOX EUROPA through Center for the Development of Industrial Technology (CDTI) within the frame of IISIS project, grant number IPT-20111023.

Acknowledgments: The tests presented in this manuscript were performed as part of the IISIS: "Investigación Integrada en Islas Sostenibles" (Integrated research in Sustainable Islands) research project, in receipt of funding from the Center for the Development of Industrial Technology (CDTI) and the Technological Fund, part of the Spanish Ministry of Industry, through the INNPRONTA research program. The final goal of the project is to research different technologies for building offshore island-cities, involving construction, energy and smart technologies with leading companies and research centers focused on each field. In the field of construction, the challenge is to develop modular construction and special marine reinforced concrete for the construction of islands that are capable of withstanding corrosive marine environments. In this context, most pathologies linked to reinforced concrete in marine environments are caused by rebar deterioration within the concrete and especially by chloride attack. The use of stainless steels is a very promising way to solve this problem, as has been demonstrated in several tests, among which those presented in this paper. Besides, the authors want to acknowledge Mario Oyarbide and Adrián Gastesi in particular for their technical support during the experimental campaign performed on these steels. Authors would like to acknowledge UPV/EHU PPGA19/61 contract as well as to the IT1314-19 (Basque Government) and GIU19/029 (UPV/EHU) research groups and to Laboratoire des ciencies de l'ingenieur appliquées, Fédération IPRA-EA4581, from the Université de Pau et Pays de l'Adour, for their support setting a cooperation framework. Finally, we are also especially thankful to ACERINOX EUROPA (part of the ACERINOX Group) for funding the IISIS project, supplying the necessary rebar samples for testing, and particularly to Rafael Sanchez and Julia Contreras from Technical Dpt./Labs for their expertise and for their commitment that greatly assisted our research.

Conflicts of Interest: The authors declare no conflict of interest.

References

1. Faccoli, M.; Roberti, R. Study of hot deformation behaviour of 2205 duplex stainless steel through hot tension tests. *J. Mater. Sci.* **2013**, *48*, 5196–5203. [CrossRef]

2. Saliba, N.; Gardner, L. Cross-section stability of lean duplex stainless steel welded I-sections. *J. Constr. Steel Res.* **2013**, *80*, 1–14. [CrossRef]
3. CEN. *Eurocode 2: Design of Concrete Structures—Part 1-1: General Rules and Rules for Buildings*; Comité Européen de Normalisation: Brussels, Belgium, 2004.
4. CEN. *Eurocode 3: Design of Steel Structures, Part 1-1: General Rules and Rules for Buildings*; Comité Européen de Normalisation: Brussels, Belgium, 2005.
5. CEN. *Eurocode 3: Design of Steel Structures, Part 1-4: General Rules-Supplementary Rules for Stainless Steels*; Comité Européen de Normalisation: Brussels, Belgium, 2006.
6. Espanya. *Código Técnico de la Edificación: (C.T.E.)*; Ministerio de Vivienda: Madrid, Spain, 2006; ISBN 978-84-340-1631-6.
7. Espanya. *Instrucción de Acero Estructural (EAE)*; Ministerio de la Presidencia: Madrid, Spain, 2011; ISBN 978-84-340-1980-5.
8. Espanya. *Instrucción de Hormigon Estructural. EHE-08*; Ministerio de Fomento: Madrid, Spain, 2011; ISBN 978-84-498-0899-9.
9. CEN. *Eurocode 3: Design of Steel Structures, Part 1-8: Design of Joints*; Comité Européen de Normalisation: Brussels, Belgium, 2005.
10. Verma, J.; Taiwade, R.V. Effect of welding processes and conditions on the microstructure, mechanical properties and corrosion resistance of duplex stainless steel weldments—A review. *J. Manuf. Process.* **2017**, *25*, 134–152. [CrossRef]
11. Paulraj, P.; Garg, R. Effect of intermetallic phases on corrosion behavior and mechanical properties of duplex stainless steel and super-duplex stainless steel. *Adv. Sci. Technol. Res. J.* **2015**, *9*, 87–105. [CrossRef]
12. Elsaady, M.A.; Khalifa, W.; Nabil, M.A.; El-Mahallawi, I.S. Effect of prolonged temperature exposure on pitting corrosion of duplex stainless steel weld joints. *Ain Shams Eng. J.* **2018**, *9*, 1407–1415. [CrossRef]
13. Vinoth Jebaraj, A.; Ajaykumar, L.; Deepak, C.R.; Aditya, K.V.V. Weldability, machinability and surfacing of commercial duplex stainless steel AISI2205 for marine applications—A recent review. *J. Adv. Res.* **2017**, *8*, 183–199. [CrossRef]
14. Kanemaru, S.; Sasaki, T.; Sato, T.; Mishima, H.; Tashiro, S.; Tanaka, M. Study for TIG–MIG hybrid welding process. *Weld. World* **2014**, *58*, 11–18. [CrossRef]
15. Kumar, M.L.S.; Verma, D.S.M.; RadhakrishnaPrasad, P. Experimental Investigation for Welding Aspects of AISI 304 & 316 by Taguchi Technique for the Process of TIG & MIG Welding. *Int. J. Eng. Trends Technol.* **2011**, *2*, 28–33.
16. Mohammed, G.R.; Ishak, M.; Aqida, S.N.; Abdulhadi, H.A. Effects of Heat Input on Microstructure, Corrosion and Mechanical Characteristics of Welded Austenitic and Duplex Stainless Steels: A Review. *Metals* **2017**, *7*, 39. [CrossRef]
17. M, M.A.; Shrikrishna, K.A.; Sathiya, P.; Goel, S. The impact of heat input on the strength, toughness, microhardness, microstructure and corrosion aspects of friction welded duplex stainless steel joints. *J. Manuf. Process.* **2015**, *18*, 92–106. [CrossRef]
18. Tasalloti, H.; Kah, P.; Martikainen, J. Effect of heat input on dissimilar welds of ultra high strength steel and duplex stainless steel: Microstructural and compositional analysis. *Mater. Charact.* **2017**, *123*, 29–41. [CrossRef]
19. Yang, Y.; Yan, B.; Li, J.; Wang, J. The effect of large heat input on the microstructure and corrosion behaviour of simulated heat affected zone in 2205 duplex stainless steel. *Corros. Sci.* **2011**, *53*, 3756–3763. [CrossRef]
20. Subramanian, G.; Palraj, S.; Balasubramanian, T.M. Galvanic corrosion interactions of zinc and SS.304 in the tropical marine atmosphere of Mandapam. *Anti-Corros. Methods Mater.* **1999**, *46*, 332–337. [CrossRef]
21. Paul, S.; Basu, K.; Mitra, P. A resistor-controlled sacrificial anode for cathodic protection of stainless steel in seawater. *Bull. Electrochem.* **2005**, *21*, 269–273.
22. ISO. *Corrosion Tests in Artificial Atmospheres—Salt Spray Tests*; International Standard Organization: Geneva, Switzerland, 2017.
23. Liu, X.; Xia, Z.; Zhou, H.; Yuan, B.; Li, Z.; Guo, F. Corrosion Behavior of Different Steel Substrates Coupled With Conductive Polymer under Different Serving Conditions. *J. Iron Steel Res. Int.* **2013**, *20*, 87–92. [CrossRef]
24. Yousefieh, M.; Shamanian, M.; Saatchi, A. Influence of Heat Input in Pulsed Current GTAW Process on Microstructure and Corrosion Resistance of Duplex Stainless Steel Welds. *J. Iron Steel Res. Int.* **2011**, *18*, 65–69. [CrossRef]

25. Sternhell, G.; Paul, D.; Taylor, D. Itzhak galvanic effects of various metallic couples on marine biofouling in a coral reef environment. *Corros. Rev.* **2002**, *20*, 453–468. [CrossRef]
26. Schaeffler, A.L. Constitution Diagram for Stainless Steel Weld Metal. *Met. Prog.* **1949**, *56*, 680.
27. ISO. *Non-Destructive Testing of Welds—Visual Testing of Fusion-Welded Joints*; International Standard Organization: Geneva, Switzerland, 2016.
28. CEN. *Non-Destructive Examination of Fusion Welds—Visual Examination*; Comité Européen de Normalisation: Brussels, Belgium, 1997.
29. CEN. *Non-Destructive Testing—Penetrant Testing—Part 1: General Principles*; Comité Européen de Normalisation: Brussels, Belgium, 1997.
30. CEN. *Non-Destructive Examination of Welds—Radiographic Examination of Welded Joints*; Comité Européen de Normalisation: Brussels, Belgium, 1997.
31. ISO. *Welding—Fusion-Welded Joints in Steel, Nickel, Titanium and Their Alloys (Beam Welding Excluded)—Quality Levels For Imperfections*; International Standard Organization: Geneva, Switzerland, 2014.
32. E04 Committee. *Test Method for Determining Volume Fraction by Systematic Manual Point Count*; ASTM International: West Conshohocken, PA, USA, 2019.
33. Calderon-Uriszar-Aldaca, I.; Briz, E.; Biezma, M.V.; Puente, I. A plain linear rule for fatigue analysis under natural loading considering the coupled fatigue and corrosion effect. *Int. J. Fatigue* **2019**, *122*, 141–151. [CrossRef]
34. Calderón-Urízar-Aldaca, I.; Briz, E.; Matanza, A.; Martin, U.; Bastidas, D.M. Corrosion Fatigue Numerical Model for Austenitic and Lean-Duplex Stainless-Steel Rebars Exposed to Marine Environments. *Metals* **2020**, *10*, 1217. [CrossRef]
35. Briz, E.; Martin, U.; Biezma, M.V.; Calderon-Uriszar-Aldaca, I.; Bastidas, D.M. Evaluation of the mechanical behavior of 2001 LDSS and 2205 DSS reinforcements exposed to simultaneous load and corrosion in chloride contained concrete pore solution. *J. Build. Eng.* **2020**, *31*, 101456. [CrossRef]
36. Calderón-Urízar-Aldaca, I.; Biezma, M.V.; Matanza, A.; Briz, E.; Bastidas, D.M. Second-order fatigue of intrinsic mean stress under random loadings. *Int. J. Fatigue* **2020**, *130*, 105257. [CrossRef]

Publisher's Note: MDPI stays neutral with regard to jurisdictional claims in published maps and institutional affiliations.

© 2020 by the authors. Licensee MDPI, Basel, Switzerland. This article is an open access article distributed under the terms and conditions of the Creative Commons Attribution (CC BY) license (http://creativecommons.org/licenses/by/4.0/).

Article

Improvement in Weldment of Dissimilar 9% CR Heat-Resistant Steels by Post-Weld Heat Treatment

Jiankun Xiong [1,2,*], Ting Li [3], Xinjian Yuan [1,3,*], Guijun Mao [2], Jianping Yang [2], Lin Yang [2] and Jian Xu [2]

1. State Key Laboratory of Long-life High Temperature Materials, Dongfang Turbine Co., Ltd., Deyang 618000, China
2. Manufacturing Technology Department, Dongfang Turbine Co., Ltd., Deyang 618000, China; 2008jiaefu@163.com (G.M.); yangjianping66@163.com (J.Y.); dtcyanglin@dongfang.com (L.Y.); xujianlut@163.com (J.X.)
3. College of Materials Science and Engineering, Chongqing University, No. 174, Shazheng Street, Shapingba District, Chongqing 400044, China; liting000@cqu.edu.cn
* Correspondence: xiongkai2010@163.com (J.X.); xinjianyuan@yahoo.com (X.Y.); Tel.: +86-23-6512-7306 (X.Y.)

Received: 6 September 2020; Accepted: 24 September 2020; Published: 2 October 2020

Abstract: The effect of the post-weld heat treatment on the microstructures and mechanical properties of the dissimilar joint of G115, a novel developed martensite heat resistant steel, and CB2 steel, currently used in an ultra-super-critical power unit, was investigated. The results indicate that the quenched martensite underwent decomposition and transformation, and the amount of dislocations were sharply decreased in the weld metal after post-weld heat treatment (PWHT). Many nano-scale $M_{23}C_6$ precipitates present in the weldment were distributed on the grain and grain boundary in a dispersed manner with PWHT. The average microhardness of the weldment decreased from about 400 HV to 265–290 HV after PWHT and only a slight decrease in the microhardness of CB2 steel was detected after PWHT at 760 °C. In contrast to the case of the as-received joint, the tensile strength of the joint was improved from 630 MPa to 694 MPa and the fracture location moved from the weld metal to the base metal after PWHT. The fracture surface consisted of a cleavage fracture mode without PWHT, whereas many dimples were observed on the fracture surface with PWHT.

Keywords: heat resistant steels; welding; post-weld heat treatment; microstructure; mechanical properties

---

1. Introduction

Heat resistant steels are widely used as structural materials for boilers, main steam pipes, and turbines of power plants [1,2]. In order to improve steam parameters and heat efficiency, researchers have turned their targets to higher-parameter thermal power units above 630 °C for developing high-capacity ultra-super-critical power plants [3]. At present, the upper limit of the operating temperature of the commercial heat-resistant steel is about 600 °C. Beyond the temperatures, problems in the still, such as thermally unstable microstructure, insufficient durability and insufficient environmental corrosion resistance, need to be considered [4]. Thus, the key factor restricting the construction of a 630 °C power station is the ability to produce heat-resistant steel in the higher temperature range of 630 °C [5,6].

In this context, many attempts have been made to overcome this limitation. A martensitic heat-resistant steel with a 9Cr-3W-3Co composition system is pivotal to research and development. MARBN (9Cr-3W-3CoVNbBN) [7–9], SAVE12AD (9Cr-3W-3CoNdVNbBN) [10], NPM (9Cr-3W-3CoVNbBN) [11], 9Cr3W (9.5Cr-3.1W-3.2CoVNbBN) [12] and G115 (9Cr-2.8W-3CoCuVNbBN) [13–16] steels have been developed. In recent years, many studies were focused on the structures and mechanical properties of G115 steel [13–16], which revealed that G115 steel has excellent structural stability and high-temperature creep properties at 630 °C. The endurance strength of G115 steel at 650 °C is 1.5 times that of P92 steel, and its resistance to high-temperature steam oxidation corrosion is better than that of P92 steel [16].

The welding research on heat-resistant steel was mostly concentrated on P91 [17,18] and P92 [19,20] steels. The joints of other steels with a 9Cr-3W-3Co composition system, such as MARBN steel, have also been reported [21,22]. As G115 steel is a newly developed heat-resistant steel, which will be used in the high temperature components; so far researches on the welding of G115 steel were limited. In this study, G115 steel was selected for welding with a heat-resistant steel (CB2), still currently used in the unit. After welding, microstructure similar to that found by casting formed in the weld bead, the heat-affected zone (HAZ) that appears adjacent to the weld line, and residual stress can be produced. To reduce these influences on the joint, post-weld heat treatment (PWHT) was employed. The improvements in the microstructure and mechanical properties of the dissimilar joints of G115 and CB2 steels with PWHT are discussed in detail in this paper.

2. Materials and Methods

G115 and CB2 steel blanks, measuring 80 × 70 × 2 mm, were machined from as-received plates and utilized as the base metals. The G115 steel plate with dimensions of 200 × 150 × 40 mm, which was produced by China BaoWu Steel Group Corporation Limited (Shanghai, China), was in the condition of quenching and high temperature tempering. A filler wire with a diameter of 2.4 mm was used and its composition was similar to that of G115 steel. The filler wire was made by Atlantic China Welding Consumables INC (Zigong, China). The chemical composition of the substrates and filler are given in Table 1.

Table 1. The chemical composition of the base metals and filler wire (wt.%).

Metals	C	Cr	W	Co	Mo	Cu	Mn	Si	Ni	V	Nb	N	B	Fe
G115	0.08	8.8	2.84	3.0	-	1.0	0.5	0.3	-	0.2	0.06	0.008	0.014	Bal.
CB2	0.11	9.18	-	0.98	1.47	-	0.69	0.3	0.33	0.2	0.06	0.02	0.009	Bal.
Filler	0.088	8.95	2.8	3.01	-	0.94	0.35	0.07	-	0.22	0.008	0.0096	0.009	Bal.

A common welding method, tungsten inert gas (TIG) welding, was used to weld G115 and CB2 steels by a TIG welding machine (YC-300WP5HGN, Panasonnic industrial machinery Co., Ltd., Tangshan, China). Based on the results of previous research, the welding current, the wire feed speed and the welding speed were 130 A, 0.5 m/min and 0.15 m/min, respectively. The post-welding heat treatment was conducted at 700, 730 and 760 °C for 2 h in a resistance heating furnace. The temperature of the solid solution of CB2 steel was about 730 °C.

Specimens for metallographic examination were sectioned from the welded joints. Ground and polished cross sections were etched in a solution (5 g $FeCl_3$ + 15 mL HCl + 80 mL H_2O) to observe the joint microstructure. Microstructural observations were conducted using an optical microscope (OM, DM2000X, Chongqing, China) and a scanning electron microscope (SEM, TESCAN VEGA 3 LMH, Brno, Czech Republic). Morphology and precipitates were analyzed by TEM. Microhardness testing was used to determine the hardness profile of the joint region. The testing was carried out using a load of 4.9 N on a hardness tester (MH_3N). The hardness testing was performed at a position 1 mm below the surface, and the distance between the hardness points was about 0.25 mm. Tensile tests were conducted by an electronic tensile test machine (AG-X 50KN, Shimadzu, Kyoto, Japan), where the loading rate was 1 mm/min. The tensile testing is shown schematically in Figure 1. Three specimens were tested for each condition. After tensile testing, the fracture surfaces of the samples were observed with SEM.

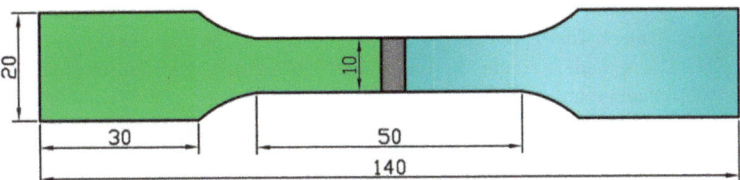

Figure 1. The schematic drawing for tensile testing (mm).

3. Results and Discussion

3.1. Microstructure of the Weld Metal

Figure 2 exhibits typical weld metal OM and SEM images of specimens prepared without and with PWHT (post-weld heat treatment) at a temperature of 730 °C after holding for 2 h. Without PWHT, the microstructure of the weld seam was dominated by lath martensite. The boundaries of martensite lath bundles were clearly visible, and the martensite in each lath bundle had relatively uniform directionality. With PWHT, the tempered martensite laths were discontinuous, and the boundaries of the martensite lath bundles became less clear. Fine grain precipitates were found on the martensite laths and the boundaries of the lath bundles; the number of precipitation phases was large and their distribution was dispersed.

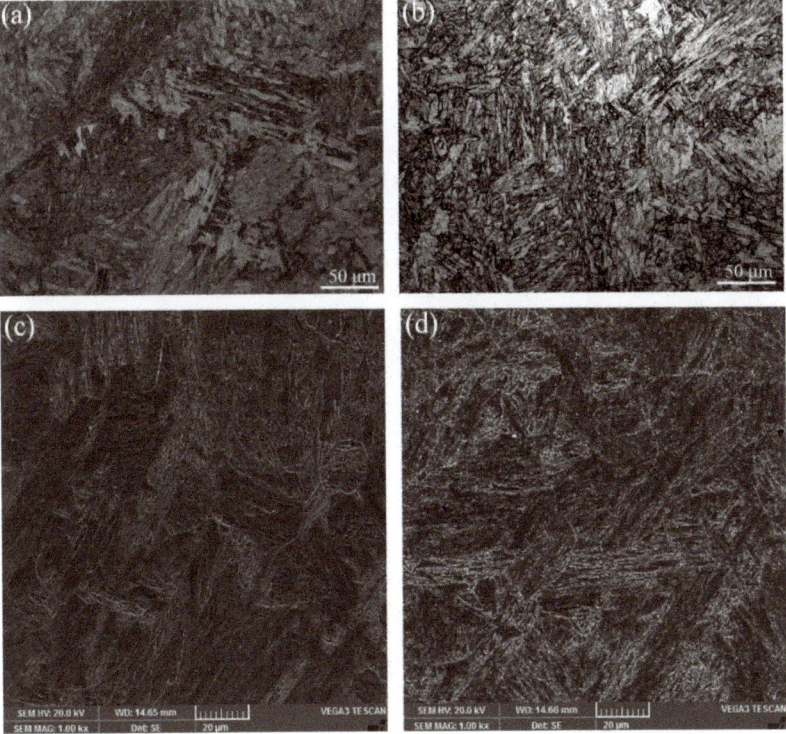

Figure 2. Optical microscope (OM) and scanning electron microscope (SEM) images of typical weld metal specimens: (**a**), (**c**) without post-weld heat treatment (PWHT); and (**b**,**d**) with PWHT at 730 °C for 2 h.

Figures 3 and 4 demonstrate the typical SEM images of HAZ (heat-affected zone) and the base metal of CB2 steel and G115 steel. The results in Figures 3a–c and 4a–c are from as-welded joints, whereas the images in Figures 3d–f and 4d–f are from the joints with PWHT at 730 °C for 2 h. The microstructure of CB2 steel and G115 steel is shown in Figure 3c,f and Figure 4c,f. Figure 3a,b, Figure 3d–e, Figure 4a,b and Figure 4d,e show steel that is in the HAZ. The positions for Figure 3a,d and Figure 4a,d were closer to the weld metal than Figure 3b,e and Figure 4b,e, respectively. In contrast with Figure 3b,e and Figure 4b,e, the size of martensite lath bundles shown in Figure 3a,d and Figure 4a,d was relatively larger, due to the higher temperature effect of these areas closer to the weld line. There are a few precipitates in the substrate of G115 steel in the joint without PWHT; that is because the as-received condition of the G115 steel is different from the condition of the CB2 steel. As compared with the as-welded joint, many precipitates were observed in the HAZ for the base metal involving CB2 and G115 steel after PWHT at 730 °C for 2 h.

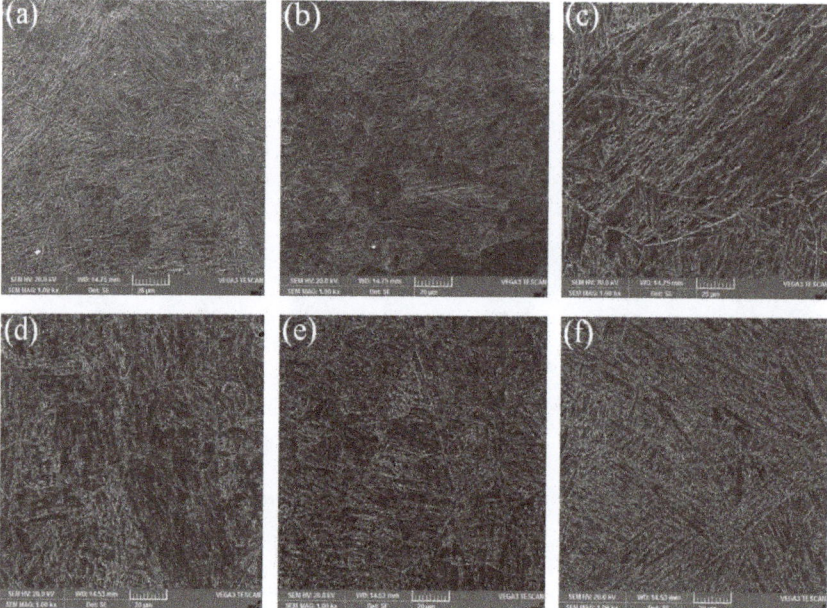

Figure 3. Typical SEM images of the heat-affected zone (HAZ) (**a**,**b**), (**d**,**e**) and base metal (**c**), (**f**) of CB2 steel: (**a**)–(**c**) without PWHT and (**d**)–(**f**) with PWHT at 730 °C for 2 h.

Figure 5 shows typical TEM (transmission electron microscope) images of the weld metal in the joints produced without PWHT. From Figure 5, the matrix phase of the weldment was martensite, and the width of the martensite lath was approximately several hundred nanometers. The distribution of the lath martensite with a certain length was regular and directional. The quenched martensite was formed by shearing and resulted in a high defect density. A large number of dislocations and dislocation tangles were detected on and between the martensite laths. These dislocations were unevenly distributed, and the dislocation cells of the cellular substructure might be formed in some high-density dislocation regions.

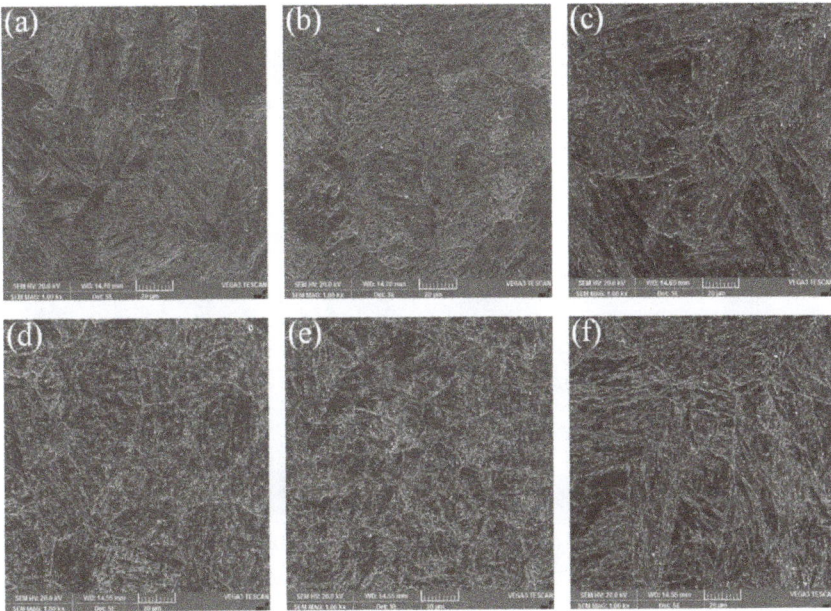

Figure 4. Typical SEM images of HAZ (**a**,**b**), (**d**,**e**) and base metal (**c**), (**f**) of G115 steel: (**a**)–(**c**) without PWHT and (**d**)–(**f**) with PWHT at 730 °C for 2 h.

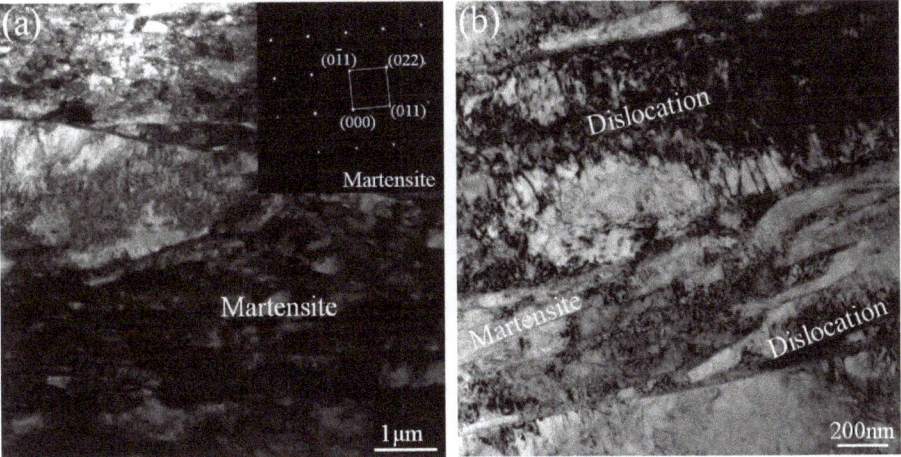

Figure 5. Typical weld metal TEM analysis of the joints produced without PWHT: (**a**) TEM images and diffraction patterns of martensite, and (**b**) dislocation.

Figure 6 illustrates typical weld metal TEM images of the joints obtained with PWHT at 730 °C for 2 h. From Figure 3, the directionality and length of the martensite laths were clearly reduced and shortened after PWHT, which can be explained by the following three reasons: (1) Firstly, the long laths of quenched martensite are decomposed, owing to the effect of the high temperature. (2) Secondly, some martensite phases undergo recovery and recrystallization processes, and then the equiaxed crystals with low dislocation density are transformed from the lath crystals. (3) Thirdly, sub-grains are formed from remaining dislocations, through multilateralization and then gradually grow up when undergoing higher temperature for an extended long time. The dislocation density was decreased significantly, which can be attributed to the reduction and disappearance of dislocations and dislocation cells in the recovery process during PWHT. The carbides were identified as $M_{23}C_6$ phase by the diffraction patterns. Many carbide particle phases were formed on the grains and grain boundaries. The precipitate size ranged from approximately several nanometers to tens of nanometers, and the size of very few carbides was approximately hundreds of nanometers.

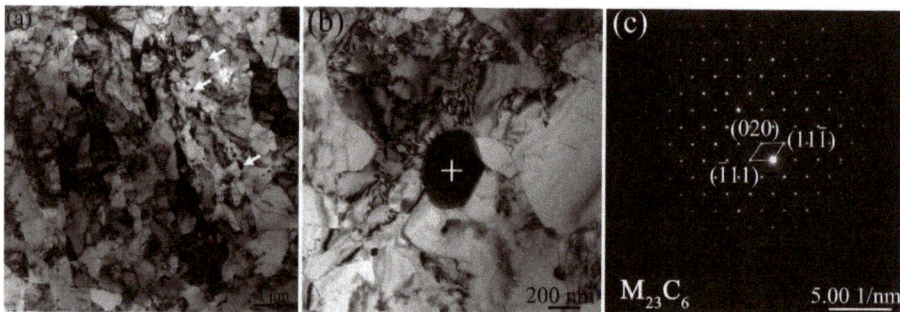

Figure 6. Typical weld metal TEM analysis of the joints obtained with PWHT at 730 °C for 2 h: (**a**) TEM images, (**b**) precipitates images, and (**c**) diffraction patterns of precipitate. The white arrows highlight the precipitates.

3.2. Microhardness of the Joint

Figure 7 displays the variation of microhardness in the joint region without and with PWHT. For the as-welded joint, the microhardness of the weld metal, whose average value is about 400 HV, is significantly higher than that of the G115 steel and CB2 steel. This might be interpreted in the following four aspects: (1) Firstly, the quenched martensite transformed by rapid cooling has higher microhardness, which results in transformation hardening in the weld zone. (2) Secondly, there are no carbide precipitates in the quenched martensite matrix, and almost all carbon atoms are dissolved into the matrix in a supersaturated form. This can lead to solid solution hardening in the weld zone. (3) Thirdly, the residual stress produced by welding and stored in the weldment makes the measured microhardness values higher. (4) Finally, a large number of dislocations (Figure 5b) are produced in the rapid cooling process and during the quenched martensite formation. This might cause dislocation hardening in the weld seam. The microhardness values of the base metal areas were higher than those of the center area, because the above-mentioned four hardening effects are greater at the faster cooling rate. The microhardness of the G115 steel was similar to that of the CB2 steel, and its value was approximately 250 HV. By comparing the two cases of without and with PWHT, an obvious change in the microhardness profile is that the microhardness of the weld seam was sharply reduced to 265–290 HV after PWHT. The transformation of tempered martensite from quenched martensite, the reduction of the solid solution carbon content in martensite, the elimination of the remaining stress and the decrease and disappearance of dislocation density during PWHT can make the microhardness of the weldment notably low. As the PWHT temperature increased, the microhardness of the G115 steel side did not change significantly, which indicates that the G115 steel has good high temperature stability. However, the microhardness of the weldment and CB2 steel side was slightly reduced especially at 760 °C. The further coarsening of grains at the higher temperature may be responsible for the decrease in microhardness.

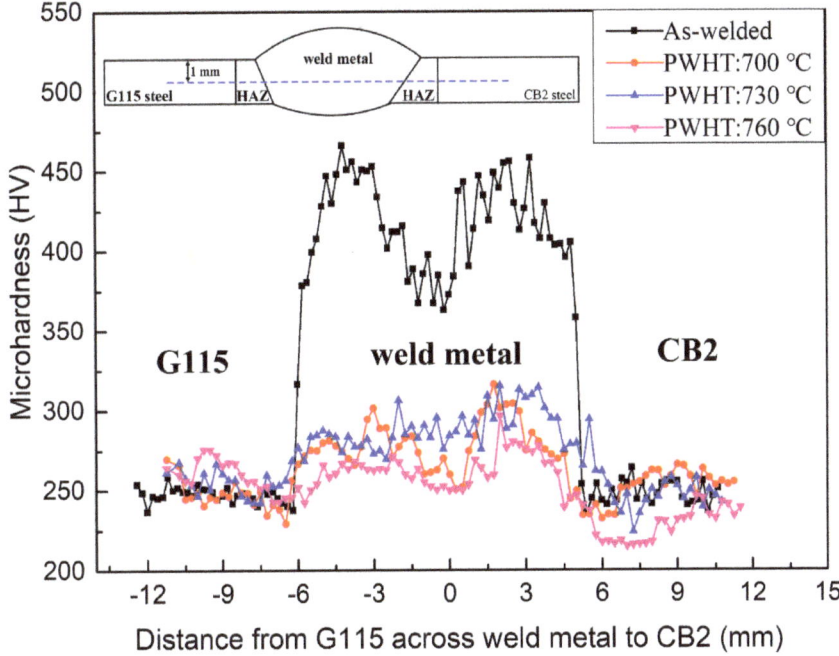

Figure 7. The microhardness profiles of the joint of G115 steel and CB2 steel without and with PWHT.

3.3. Tensile Strength of the Joint

Figure 8 presents the tensile strengths of the joints without and with PWHT. The tensile strength of the joint without PWHT was about 630 MPa, and the fracture during tensile testing was located in the weld seam. Due to the large amount of fine carbides with dispersed distribution (Figure 6), the tensile strength values of the joints that underwent PWHT were notedly larger than those without PWHT, and the failure during tensile testing took place in the base metal of CB2, which indicates that the thermal strength of CB2 is lower than that of G115. As the PWHT temperature increased from 700 to 730 °C, the joint strength improved markedly, whereas the joint strength clearly decreased by further increasing the PWHT temperature from 730 to 760 °C. The reduction in the properties including microhardness and tensile strength of CB2 steel at 760 °C might be because the temperature was higher than the solid solution temperature of CB2 steel. The maximum strength of the joint that was reached was 694 MPa at the PWHT temperature of 730 °C, which was approximately 94 % of the tensile strength of the base metal (Table 2).

Table 3 shows the yield strength and elongation to failure of the welded joints without and with PWHT. The yield strength values of the joints with PWHT are much larger than those without PWHT. The yield stress exhibits a change that firstly increases and then decreases as the temperature of PWHT enhances, which was similar to the change in the tensile strength. The maximum yield strength obtained with PWHT at 730 °C was higher than that of the joint without PWHT by about 159 MPa. The ductility of the joint increased significantly after PWHT, and the elongation values of the joints with PWHT were an order of magnitude larger than those without PWHT.

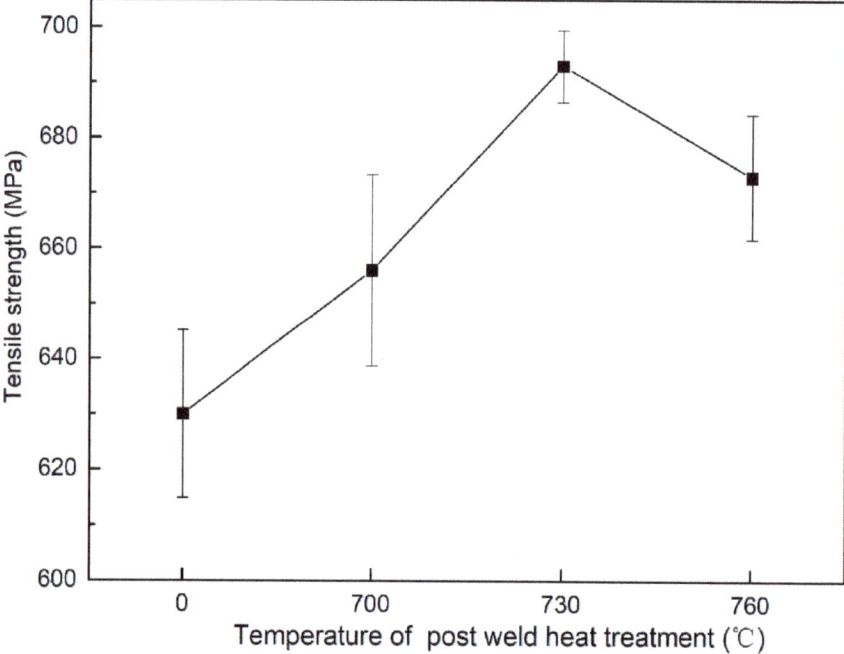

Figure 8. Tensile strengths of the joints of G115 steel and CB2 steel without and with PWHT.

Table 2. Elongation, yield strength and tensile strength of the base metals.

	As-Received G115	As-Received CB2	G115 at 730 °C for 2 h	CB2 at 730 °C for 2 h
Elongation (%)	8.92	6.14	10.6	7.79
Yield strength (MPa)	641	623	621	614
Tensile strength (MPa)	749	746	745	740

Table 3. Elongation and yield strength of the welded joints without and with PWHT.

	As-Weld Joint	PWHT at 700 °C for 2 h	PWHT at 730 °C for 2 h	PWHT at 760 °C for 2 h
Elongation (%)	0.58	6.66	6.69	7.12
Yield strength (MPa)	410	547	569	539

3.4. Fracture Characteristics

Figure 9 represents the fracture surface images of the joints without and with PWHT of 730 °C. The fracture of the as-welded joint occurred in the weldment; the failed surface contained cleavage fracture characteristics and it is considered to be a mainly brittle fracture. In all cases investigated for PWHT, the weld seam was strengthened and CB2 steel became a weak part, causing fractures that occurred in the CB2 side. The post fracture microstructures of the joints were similar and there were many dimples on the fracture surface. The results indicate that the PWHT process improved the microstructure and mechanical properties of the weldment, thereby avoiding brittle fracture at the weld.

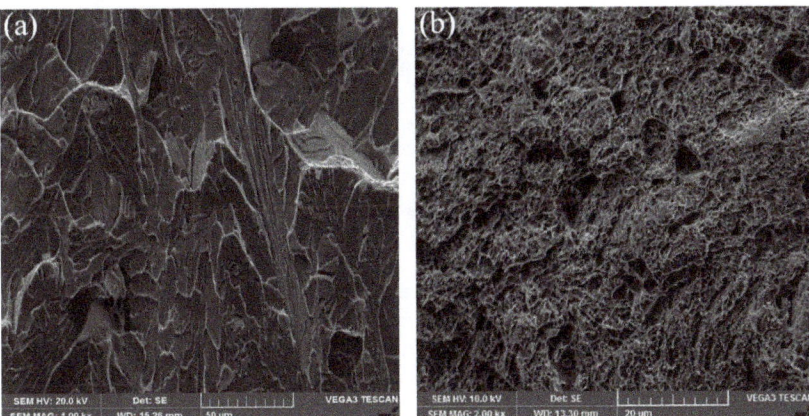

Figure 9. Typical fracture surface SEM images of the joints: (**a**) without PWHT and (**b**) with PWHT, at 730 °C for 2h.

4. Conclusions

In this paper, the microstructures and mechanical properties of the weldment between G115 and CB2 steels without and with PWHT were studied. The related conclusions can be drawn as follows:

(1) Without PWHT, the weld seam was mainly composed of the quenched martensite with a lath thickness of approximately several hundred nanometers Many dislocations and dislocation tangles were found on and between the martensite laths. With PWHT, the tempered martensite was transformed from the quenched martensite and the number of dislocations was drastically reduced. In addition, $M_{23}C_6$ secondary-phases with the size of approximately several nanometers to tens of nanometers were dispersedly distributed on the grains and grain boundaries.

(2) Without PWHT, the average microhardness of the weldment (~400 HV) was evidently larger than that of the G115 steel and CB2 steel (~250 HV). With PWHT, the microhardness of the weldment decreased to 265–290 HV and the microhardness of CB2 steel was slightly reduced at 760 °C for PWHT.

(3) Without PWHT, the tensile strength of the joint was about 630 MPa, and the fracture with mainly cleavage fracture characteristics occurred in the weld seam. With PWHT, the joint strength increased significantly and reached up to a maximum value (694 MPa) at 730 °C, which might result from many dispersed precipitates, and the fracture with many dimples was located in the base metal of CB2 steel.

Author Contributions: J.X. (Jiankun Xiong), X.Y., J.Y., L.Y., and J.X. (Jian Xu) formulated the original problem, designed the study, developed the methodology, and wrote the manuscript. J.X. (Jiankun Xiong), T.L., X.Y., G.M. and J.X. (Jian Xu) performed the welding, testing, characterization and collected data. J.X. (Jiankun Xiong) and G.M. controlled the materials selection. J.X. (Jiankun Xiong) and X.Y. provided guidance and interpretation of data. All authors have read and agreed to the published version of the manuscript.

Funding: This work was financially supported by the Sichuan Deyang Open University-City Cooperative Technology Research and Development Project (Project No.: 2018CKJ004) and the Sichuan Deyang Industry-University-Research Cooperation Technology R&D Project (Project No.: 2019CK094). This research was also supported by the fund of Sate Key Laboratory of Long-life High Temperature Materials (Project No.: DTCC28EE190934).

Conflicts of Interest: The authors declare no conflict of interest.

References

1. Fedoseeva, A.; Dudova, N.; Kaibyshev, R.; Belyakov, A. Effect of tungsten on creep behavior of 9%Cr–3%Co martensitic steels. *Metals* **2017**, *7*, 573. [CrossRef]

2. Kaybyshev, R.O.; Skorobogatykh, V.N.; Shchenkova, I.A. New martensitic steels for fossil power plant: Creep resistance. *Phys. Met. Metallogr.* **2010**, *109*, 186–200. [CrossRef]
3. Wang, X.; Wang, X.; Luo, B.; Hu, X.; Yuan, T. Creep degradation assessment in 9%Cr heat-resistant steel welded joints using ultrasonic methods. *Results Phys.* **2019**, *12*, 307–320. [CrossRef]
4. Zinkle, S.J.; Was, G.S. Materials challenges in nuclear energy. *Acta Mater.* **2013**, *61*, 735–758. [CrossRef]
5. Rojas, D.; Garcia, J.; Prat, O.; Sauthoff, G.; Kaysser-Pyzalla, A.R. 9%Cr heat resistant steels: Alloy design, microstructure evolution and creep response at 650 °C. *Mater. Sci. Eng. A* **2011**, *528*, 5164–5176. [CrossRef]
6. Wang, H.; Yan, W.; Zwaag, S.V.; Shi, Q.; Wang, W.; Yang, K.; Shan, Y. On the 650 °C thermostability of 9e12Cr heat resistant steels containing different precipitates. *Acta Mater.* **2017**, *134*, 143–154. [CrossRef]
7. Abe, F. New Martensitic Steels. In *Materials for Ultra-Supercritical and Advanced Untra-Supercritical Power Plants*; Gianfrancesco, A.D., Ed.; Woodhead Publishing: Cambridge, UK, 2017; pp. 323–374.
8. Abe, F. Development of Creep-Resistant Steels and Alloys for Use in Power Plants. In *Structural Alloys in Power Plants*; Shirzadi, A., Jackson, S., Eds.; Woodhead Publishing: Cambridge, UK, 2014; pp. 250–293.
9. Abstossa, K.G.; Schmigallab, S.; Schultzeb, S.; Mayr, P. Microstructural changes during creep and aging of a heat resistant MARBN steel and their effect on the electrochemical behaviour. *Mater. Sci. Eng. A* **2019**, *743*, 233–242. [CrossRef]
10. Iseda, A.; Yoshizawa, M.; Okada, H.; Hamaguchi, T.; Hirata, H.; Joutoku, K.; Ono, T.; Tanaka, K. Development of 9Cr ferritic steel tube and pipe SAVE12AD for advanced power boilers. *Therm. Nucl. Power Gener. Conv. Collect. Works* **2016**, *12*, 49–55.
11. Hollner, S.; Piozin, E.; Mayr, P.; Caës, C.; Tournié, I.; Pineau, A.; Fournier, B. Characterization of a boron alloyed 9Cr3W3CoVNbBN steel and further improvement of its high-temperature mechanical properties by thermomechanical treatments. *J. Nucl. Mater.* **2013**, *441*, 15–23. [CrossRef]
12. Fedoseeva, A.; Dudova, N.; Kaibyshev, R. Creep behavior and microstructure of a 9Cr–3Co–3W martensitic steel. *J. Mater. Sci.* **2017**, *52*, 2974–2988. [CrossRef]
13. Xiao, B.; Xu, L.; Zhao, L.; Jing, H.; Han, Y.; Zhang, Y. Creep properties, creep deformation behavior, and microstructural evolution of 9Cr-3W-3Co-1CuVNbB martensite ferritic steel. *Mater. Sci. Eng. A* **2018**, *711*, 434–447. [CrossRef]
14. Liu, Z.; Liu, Z.; Wang, X.; Chen, Z. Investigation of the microstructure and strength in G115 steel with the different concentration of tungsten during creep test. *Mater. Charact.* **2019**, *149*, 95–104. [CrossRef]
15. Xiao, B.; Xu, L.; Cayronc, C.; Xue, J.; Sha, G.; Logé, R. Solute-dislocation interactions and creep-enhanced Cu precipitation in a novel ferritic-martensitic steel. *Acta Mater.* **2020**, *195*, 199–208. [CrossRef]
16. Yu, Y.; Liu, Z.; Zhang, C.; Fan, Z.; Chen, Z.; Bao, H.; Chen, H.; Yang, Z. Correlation of creep fracture lifetime with microstructure evolution and cavity behaviors in G115 martensitic heat-resistant steel. *Mater. Sci. Eng. A* **2020**, *788*, 139468. [CrossRef]
17. Silva, F.J.G.; Pinho, A.P.; Pereira, A.B.; Paiva, O.C. Evaluation of welded joints in P91 steel under different heat-treatment conditions. *Metals* **2020**, *10*, 99. [CrossRef]
18. Wang, Y.; Li, L.; Kannan, R. Transition from type IV to type I cracking in heat-treated grade 91 steel weldments. *Mater. Sci. Eng. A* **2018**, *714*, 1–13. [CrossRef]
19. Wang, X.; Pan, Q.; Liu, Z.; Zeng, H.; Tao, Y. Creep rupture behaviour of P92 steel weldment. *Eng. Fail. Anal.* **2011**, *18*, 186–191.
20. Sklenička, V.; Kuchařová, K.; Svobodová, M.; Kvapilová, M.; Král, P.; Horváth, L. Creep properties in similar weld joint of a thick-walled P92 steel pipe. *Mater. Charact.* **2016**, *119*, 1–12. [CrossRef]
21. Mayr, P.; Martín, F.M.; Albu, M.; Cerjak, H. Correlation of creep strength and microstructural evolution of a boron alloyed 9Cr3W3CoVNb steel in as-received and welded condition. *Mater. High Temp.* **2014**, *27*, 67–72. [CrossRef]
22. Matsunaga, T.; Hongo, H.; Tabuchi, M.; Sahara, R. Suppression of grain refinement in heat-affected zone of 9Cr-3W-3Co-VNb steels. *Mater. Sci. Eng. A* **2016**, *655*, 168–174. [CrossRef]

© 2020 by the authors. Licensee MDPI, Basel, Switzerland. This article is an open access article distributed under the terms and conditions of the Creative Commons Attribution (CC BY) license (http://creativecommons.org/licenses/by/4.0/).

Article

Microstructure and Performance Analysis of Welded Joint of Spray-Deposited 2195 Al-Cu-Li Alloy Using GTAW

Chuanguang Luo [1,2], Huan Li [1], Yonglun Song [3,*], Lijun Yang [1,*] and Yuanhua Wen [2]

1. Tianjin Key Laboratory of Advanced Joining Technology, Tianjin University, Tianjin 300072, China; chg_luo@163.com (C.L.); lihuan@tju.edu.cn (H.L.)
2. Sichuan Aerospace Changzheng Equipment Manufacturing Co., Ltd., Chengdu 610100, China; wyh20017@163.com
3. College of Mechanical Engineering and Applied Electronics Technology, Beijing University of Technology, Beijing 100124, China
* Correspondence: 13911138686@163.com (Y.S.); yljabc@tju.edu.cn (L.Y.); Tel.: +86-13911138686 (Y.S.); +86-13002244217 (L.Y.)

Received: 26 July 2020; Accepted: 9 September 2020; Published: 14 September 2020

Abstract: High-strength aluminum alloy fabricated using spray deposition technology possesses many advantages, such as fine crystal grains, low component segregation, uniform microstructure, and small internal stress. In this study, spray-deposited 2195 Al-Cu-Li alloy in forged state was used and welded using the gas tungsten arc welding (GTAW) process to test and verify the features of the fusion joint. Quantitative analysis was carried out to evaluate the relationship between the local microstructures and performances of the fusion joint, which was composed of four zones: weld metal, fusion zone, heat-affected zone, and base metal. The characteristic quantities of each zone, including recrystallized grain fraction, grain sizes, grain misorientation angle, and Vickers hardness, and their distributions were considered as the key factors affecting the performance of the joint because of welding thermal cycle impact on the fusion joint. To recognize the metallurgical characteristics of spray-deposited alloy 2195, a statistical algorithm based on the concept of the Hall–Petch relationship was proposed to validate the actual test results, which include the correlation effects of both the filler wire and welding process. The correlation between the microstructures and performances of several characteristic quantities were evaluated by integrating the above characteristic information of the fusion joint under the strong coupling of multiple factors. Thus, the advantages of weldability of spray-deposited alloy 2195 using GTAW could be understood in detail.

Keywords: spray-deposited Al-Cu-Li alloy; GTAW; microstructure; welded joint; performance analysis

---

1. Introduction

Al-Cu-Li alloy is a new high-strength material that possesses low density and high elasticity. It is in increasing demand by the manufacturers of structural parts for aerospace applications. Alloy 2195, which belongs to the Al-Cu-Li-Ag-Mg series, has been used in lightweight fuel tank structural parts of aerospace vehicles [1–5]. In this application, the cylinder and vessel head are welded with bending rolled plates after spin forming, whereas the connectors such as flanges and connecting rings formed by forging are assembled and welded to these parts. Fabrication of these parts through traditional melt casting and forging often result in defects such as component segregation, oxide inclusion, delayed cracking, and large residual stress. Furthermore, joint cracks in welded forging flanges occur frequently. Cracking occurs mostly on the side of the fusion zone and is closely related to the metallurgical defects of forging materials. These problems have been proved to be detrimental to aerospace vehicle tank structures [6].

High-strength Al alloy fabricated using spray deposition possesses many advantages [7–10], such as fine crystal grains, low component segregation, uniform structure, and small internal stress, as well as short production cycle. Spray deposition is performed under vacuum or an inert gas protection atmosphere to ensure that the work pieces have low oxygen content and less oxide inclusions, and it is conducive to subsequent thermal processing. Furthermore, the machining allowance of forgings and annular parts can be significantly reduced.

In this study, forged plates of spray-deposited alloy 2195 were welded using the gas tungsten arc welding (GTAW) process to test and observe the characteristics of the fusion joint. Because of the impact of the welding thermal cycle on the fusion joint, it was composed of four zones: weld metal (WM), fusion zone (FZ), heat-affected zone (HAZ), and base metal (BM). The characteristic quantities of each zone, including the recrystallized grain fraction, grain sizes, grain misorientation angle, and Vickers hardness, and their distributions were considered as the key factors affecting the performance of the joint because of the impact of the welding thermal cycle on the fusion joint. To recognize the metallurgical characteristics of spray-deposited alloy 2195, a statistical algorithm based on the concept of the Hall–Petch relationship was proposed to validate the actual test results, which include the correlation effects of both the filler wire and welding process. The correlation between the microstructures and performances of several characteristic quantities was evaluated by integrating the above characteristic information of the fusion joint under the strong coupling of multiple factors. Thus, the advantages of weldability of spray-deposited alloy 2195 using GTAW could be understood in detail.

2. Experimental Procedure

2.1. Materials

In this study, cylindrical blanks of the spray-deposited alloy 2195 (designation: D2195; manufacturer: Jiangsu HR Spray Forming Alloy Co., Ltd., Zhenjiang, China) were Φ 300 mm in diameter. The blanks were cogged and forged into pieces (size: 300 mm × 300 mm × 60 mm), sequentially annealed at 425 °C for 2 h, cooled in the furnace for 35–40 min, and then cooled in air. The solution was treated at 505 °C for 70 min in a protection atmosphere furnace, with the time of quenching transfer to water being less than 3 s, and quenched in water at not more than 20 °C, and the cooling rate was approximately 120 °C/s. This is followed by artificial aging at 165 °C for 40 h in the T6 (solution + artificial aging) state. This heat treatment system was specifically developed independently for alloy D2195. The test plates were processed by milling to a size of 300 mm (L) × 150 mm (W) × 6.0 mm (H). The chemical composition of alloy D2195 is listed in Table 1.

Table 1. Chemical composition of D2195 (wt.%).

Component	Cu	Mn	Mg	Ag	Si	Fe	Zr	Li	Al
D2195	3.80	0.0006	0.45	0.30	0.066	0.035	0.12	0.86	Bal.

The microstructure characteristics of alloy D2195 forged specimens after the T6 treatment were obtained using electron backscatter diffraction (EBSD). The microstructure and misorientation angle distribution are depicted in Figure 1. The mechanical properties and grain sizes of the T6 treatment-forged specimens are presented in Table 2.

Table 2. Mechanical properties and grain size of forged plate D2195-T6.

UTS σ_b MPa	YS $\sigma_{0.2}$ MPa	EL %	Average Grain Size/ Standard Deviation μm	Misorientation Angle (2–15°) %	Average Vickers Hardness $H_{V0.5}$
495	422	9.5	10.3/3.9	31.0	156

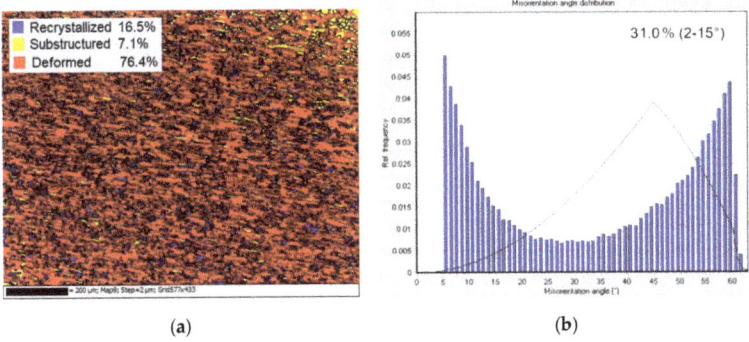

Figure 1. (a) Microstructure and (b) misorientation grain distribution of forged D2195-T6.

2.2. Welding Procedure

Welding test plates were made by manual GTWA process with two-layer welds. The welding conditions and relevant parameters are listed in Table 3. The welding parameters, preheating temperature, joint form, welding sequences, heat input, and other factors were adjusted before the actual welding test to optimize reasonable thermal cycles, to reduce or eliminate various metallurgical defects, and to improve the state of joint microstructure. All the welded test plates were examined using radiographed detection in accordance with the standard of space containers.

Table 3. Welding conditions and main process parameters.

Parameter	Value
Surroundings	Temperature: 22–24 °C, Relative humidity: 40–60%
Welding power source Operation W-Zr electrode	MILLER Dynasty-700, AC Pulse-TIG, Manual welding with two layers Φ 4 mm, 60° conical fillet
Shielding gas Flow rate of shielding gas Back protective gas Preheating Groove type Penetration pass Welding speed Filler pass Welding speed Filler wire	Pure Ar (99.999%) 16 L/min Pure Ar (99.999%), 25 L/min 80–90 °C V-95° with 1.5 mm blunt edge 200–220 A/16.5–17.5 V 130–140 mm/min 180–200 A/17.5–18.5 V 90–100 mm/min Φ 3.2 mm (details in Table 4)

Table 4. Chemical compositions of filler wire and WM of joint (wt.%).

Component	Cu	Ag	Zr	Ti	Si	Al
Wire	6.44	0.35	0.26	0.35	0.28	Bal.
WM	4.15	0.30	0.20	0.12	0.12	Bal.

To accurately understand the thermal input and thermal cycle of the test plate during the welding process, slots with a depth of 1 mm were marked on the back side of the test plate. These slots were at 4 mm, 6 mm, and 8 mm distances from the center of the weld, as depicted in Figure 2. K-type thermocouples were used in the test, and the measurement range was 0–800 °C with a measurement error of ±5 °C. The thermocouples were calibrated before the test to eliminate the effect of "temperature drift" on the measurement results. The welding thermal cycle curves were recorded to obtain their corresponding temperature histories of the filler pass and analyze the evolution of the joint microstructure. The temperature experience and corresponding microstructural impact in each zone are also reflected in Figure 3.

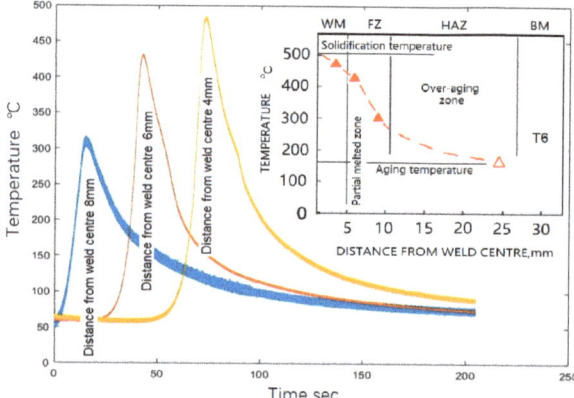

Figure 2. Recorded thermal cycle curves in welding and the local zone conditions.

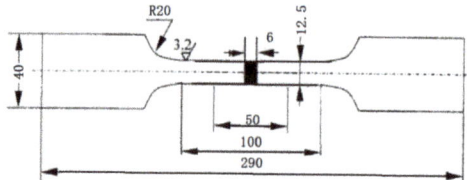

Figure 3. Dimensions of tensile test specimen (in mm).

The test plate and welding fixture were preheated prior to welding to reduce the heat-induced stress and deformation of the test plate.

2.3. Analysis of the Fusion Joint

A special filler wire was designed and developed after many tests for the fusion welding of alloy D2195. The chemical compositions of the filler wire and WM of the joint are listed in Table 4. After welding, a test plate was fabricated to the size specified by the Chinese Standards GB/T 16865-2013 and GB/T2654-2008, with a gauge of 50 mm, as depicted in Figure 3.

To obtain the performance of the fusion joint, tensile and Vickers hardness tests were performed. The fracture location of a typical tensile specimen is shown in Figure 4. The distance from the center of the weld was approximately 15 mm. The Vickers hardness was measured at those locations in the cross section of each joint, which were 1.5 mm and 4.5 mm away from the bottom edge of each test plate, in accordance with the requirements of the "Hardness Test Methods on Welded Joints" (GB/T 16865). Two adjacent points were spaced at 0.5 mm from the center of each joint to the BM. The test loading force was 500 gf and the retention load time was 10 s. The location for the hardness tests is depicted in Figure 5a, and the metallographs and morphologies of the WM and each local zone are depicted in Figures 5b and 5c, respectively.

Figure 4. Fracture location of the tensile specimen.

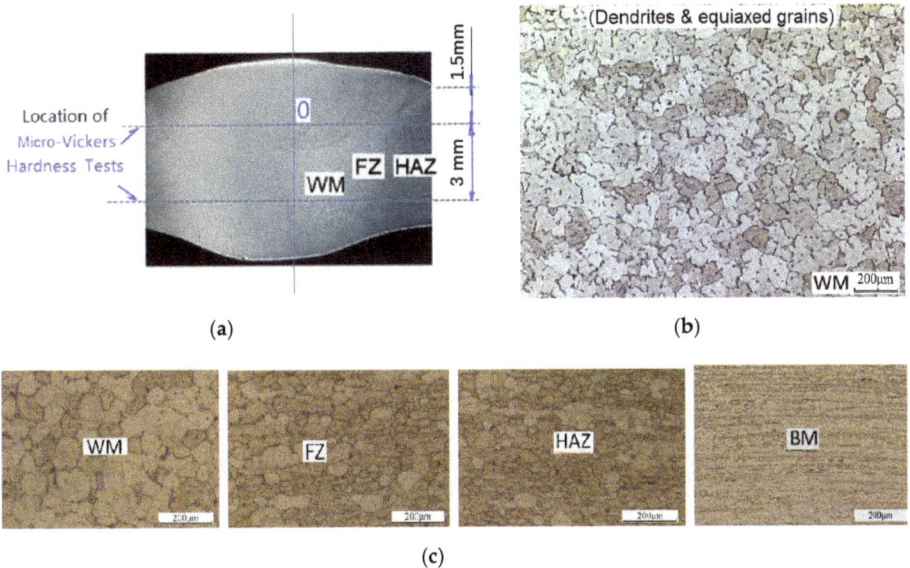

Figure 5. Metallographs and morphologies of the joint with each local zone. (**a**) Location for the hardness tests; (**b**) Microstructure of the WM; (**c**) Metallographs and morphologies of the WM and each local zone.

Figure 6 shows the Vickers hardness distributions from the WM to the BM. All the specimens were welded under the same welding condition as shown in Table 3; samples 1 and 2 are repetitions. The microstructural characteristic quantities of WM, FZ, HAZ, and BM were obtained using EBSD, as depicted in Figure 7. Subsequently, the statistics of the recrystallized grain fraction, grain sizes, misorientation angles, and their distributions were obtained and the corresponding microstructure morphologies were observed and contrasted. In general, the grain size in each local zone of the joint appeared fine and uniform, which is favorable for the suppression or elimination of various types of microcracks during welding.

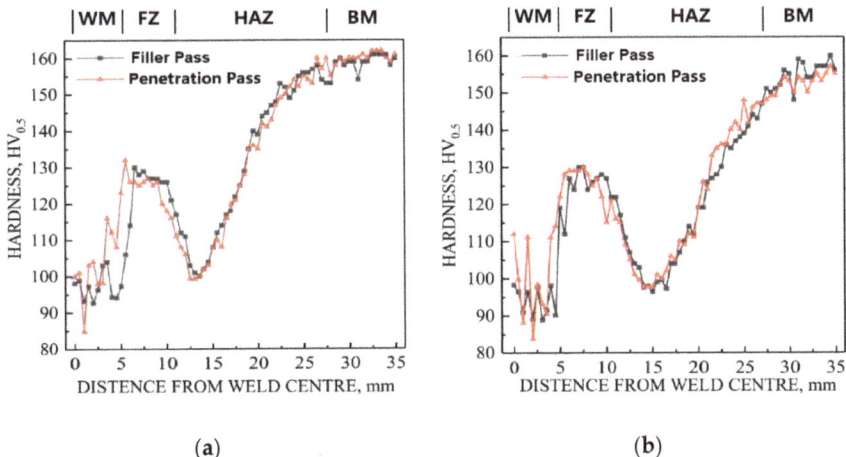

Figure 6. The Vickers hardness distributions from WM to BM: (**a**) sample 1 and (**b**) sample 2.

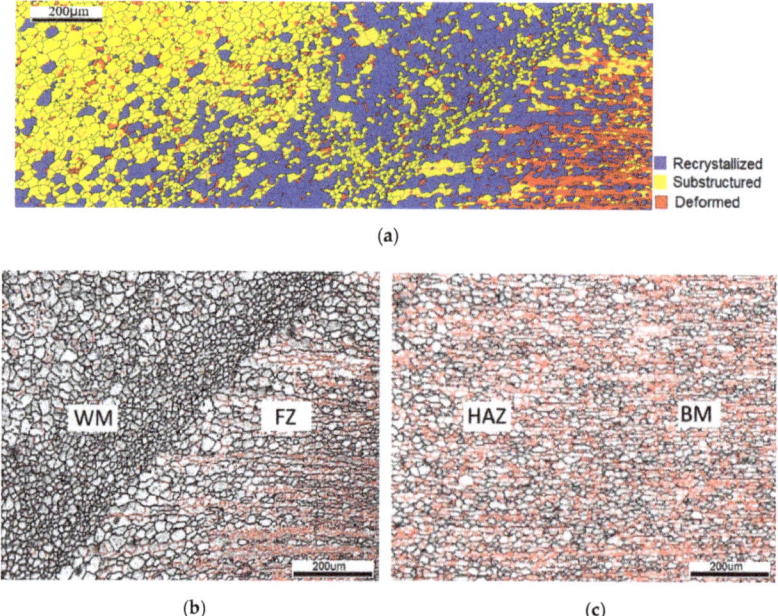

Figure 7. (a) Microstructure characteristic quantities of (b) joint and (c) grain morphologies.

The corresponding temperature histories of thermal cycles at three locations of the joint hardness distributions and microstructure morphologies of various zones indicate the following:

(1) The WM was located at the range of approximately 5 mm from the center of the weld to each side. Because of the V-95°-shape groove, the widths of the front and back joints of the test plate were approximately 10 mm and 8 mm, respectively. While the filler pass was welded, the thermal cycle history of the weld metal at a distance of 4 mm from the weld center can be described as follows: the instantaneous peak temperature was approximately 480 °C and the retention time was approximately 40 s in the range of 180–200 °C. The crystallizing morphology on the WM was dendrites and equiaxed grains with a high substructure grain fraction, and the strengthening element of filler wire compositions formed a limited solid solution effect during the thermal cycle. In addition, dispersive and fine aluminized substance points such as Al_3Zr and Al_3Ti can exist in the WM owing to the peritectic reaction of Zr and Ti in the filler wire (refer to Table 4) during the molten pool crystallization to bring about the effect of refining grains, whose microscopic morphology is shown in Figure 7b.

(2) The weld fusion line is an interface between the WM and HAZ in the fusion joint. In general, the microstructure of the FZ including the fusion line and its two sides would be more complex than that of the other zones of the joint. Under the experimental conditions, the FZ was 5.5–11 mm from the weld center. Because the V-shaped groove had a 1.5 mm blunt edge, there was a dilution rate of 15–20% between the filler wire and BM on the chemical compositions (refer to Table 4) at both sides of the fusion line, so that alloy elements may be interpenetrated and the grain boundary segregation phenomena may occur. Therefore, aspects such as the microstructure and performance of the joint changed significantly in terms of the hardness distribution (Figure 6). The measured instantaneous temperature peaked at approximately 430 °C during a thermal cycle at a location 6 mm from the weld center, and the retention time was approximately 40 s at approximately 150 °C. The recrystallization fraction of the microstructures peaked in this zone. In addition, a local "strengthening" phenomenon appeared because of the climbing and rearrangement of dislocations during the formation of solidified substructures. The micromorphology of the fusion zone is shown in Figure 7a.

(3) The HAZ of the joint was 11.5–27.5 mm from the weld center, whereas the instantaneous temperature peak was at approximately 300 °C and the retention time was approximately 150 °C to result in the BM partially dissolved and produce a softening zone which was about 16 mm wide. Its microstructure retained the originally highly deformed grain fraction of the BM. The metallography of the joint zone is depicted in Figure 7b.

(4) The deformed grain fraction of BM was up to 76.4%, which indicates that significantly high internal energy was stored in the base metal during the forming processes to affect the microstructure of the joint and the evolution of the grain.

3. Analysis Method

3.1. Establishment of an Adaptation Model Based on the Hall–Petch Relation

Understanding the correlation between the microstructure and performance of high-strength Al alloy welded joints is of significance for regulating and optimizing welding processes. Therefore, a quantitative analysis of the process has been a research area since a long time. Generally, the Hall–Petch relationship has been applied for the analysis of the relationship between material microstructures and performances [11,12]. The relationship between the material yield strength (σ_y) and its average grain diameter (d) can be expressed as

$$\sigma_y = \sigma_0 + k_y d^{-1/2} \tag{1}$$

where σ_0 is the friction stress and k_y is a positive yielding constant associated with the stress required to extend the dislocation activity into adjacent unyielded grains. Equation (1) can be reformulated in terms of the hardness, H_v, through the following relation [13,14]:

$$H_v = H_0 + k_H d^{-1/2} \tag{2}$$

where H_0 and k_H are the appropriate constants associated with the hardness measurements. Because σ_0, H_0, and k are chemistry- and microstructure-dependent constants, metallurgical factors such as melting, semimelting, overaging, dissolution softening, and grain coarsening would occur frequently in fusion joints of solid solution strengthened aluminum alloys. It would be difficult to fully express the correlation between the microstructure and performance of the joints of such Al alloys using only characteristic quantities such as the grain size and microhardness. As depicted in Figure 8a, even though a trend in the Hall–Petch relationship is observed, it reflects the effects of multiple microstructure factors on the behavior of the performance [5,15,16]. Some characteristic information that characterizes the joint microstructural evolution should be considered and selected in accordance with the characteristics of the solid solution strengthened Al alloy in the case of fusion welded joints to further participate in and integrate into the analysis of the correlation of joint performances. Therefore, the distributions of the recrystallized grain fraction and misorientation angle of each zone were considered as additional factors; the features are depicted in Figure 8b.

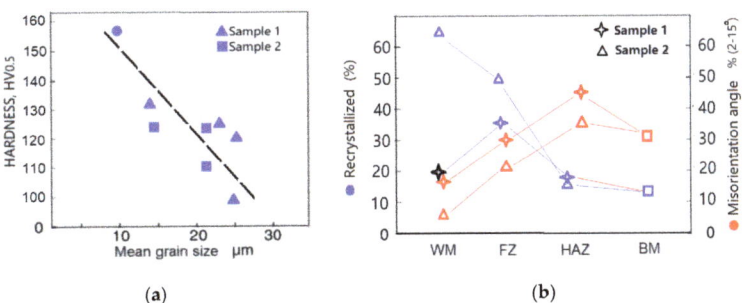

Figure 8. (a) Trend in H–P relationship and (b) distributions of recrystallized grain fraction and misorientation angle of each zone of joints.

Based on the actual measurements of physical samples under certain welding conditions, a corresponding phenomenological statistical model of the evolution of the microstructure has been proposed to evaluate or predict the local performance evolution of welded joint, which is an available research method for thermoforming manufacturing including welding. The analysis of the correlation between the microstructure and performance of a welded joint not only depends on its chemical composition but also relates to the welding thermal cycle-induced microstructure evolution and multiple factors of different attributes, such as the welding process parameters. Therefore, both characteristic quantities, namely, joint recrystallization grain fraction affected directly by the welding thermal cycle and the misorientation angle distribution related to the material mechanical plastic deformation and dislocation climbing behaviors, should be chiefly taken into account based on the statistical analysis model of a few characteristic quantities. The former is related to the microstructure states of WM and FZ, whereas the latter relates to their comprehensive mechanical properties. Additionally, the correlation between the joint microstructure and its performance can be reflected more comprehensively by integrating statistical distributions of the grain size and Vickers hardness of each zone. Therefore, a statistical analysis model for correlation between some characteristic quantities of various attributes and joint performance is proposed.

Based on the multidimensional vector inner product principle of the similar system theory and multiattribute characteristic information, the degree of difference in its multifactor fit (F–fit) was defined as the distance vector $(d_{j,R})$, which is also described as the difference between the statistical value (a_j) and its corresponding reference quantity (b_R) in n characteristic factors. It is equivalent to the distance vector sum $(d_{j,R})$ of various characteristic factors $(a_{j,i})$ and their corresponding reference quantities $(b_{R,i})$. It can be expressed as:

$$d_{j,R} = \|a_j - b_R\| = \sqrt{\sum_{i=1}^{n}(a_{j,i} - b_{R,i})^2} \tag{3}$$

The statistical significance of Equation (3) is the standard deviation of the degree of difference between n characteristic factors $(a_{j,i})$ and their corresponding reference quantities $(b_{R,i})$. The continuity of the evolution of joint microstructures is a prerequisite for the selection of a statistical reference, which is the interface of each zone in the metallurgical sense. Therefore, the FZ of a joint is determined as the adaptation reference of its WM and HAZ to conform to the above physical meaning. Similarly, the HAZ serves as an adaptation reference to the BM.

3.2. Calculation of the Fittingness of a Joint Based on Some Characteristic Quantities

Table 5 lists the statistical values that characterize the correlation between structures and performances of each zone of welded joints. Two samples were selected based on the welding conditions (Table 3). Because manual welding may result in some fluctuations in welding, the characteristic quantities (such as the misorientation angle distribution and recrystallized grain fraction) of microstructures of the both samples fluctuated to a certain extent; however, their grain sizes and microhardness were basically not influenced. These cases also reflect the random effects of actual welding conditions on the sample microstructure and performance.

Table 5. Statistical characteristic values of the joint samples.

Distances from Weld Centre (mm) // Local Zone	Recrystallize Fraction %	Misorientation Angle (<15°) %	Average Grain Size μm	Average Vickers Hardness $H_{V0.5}$
0–5//WM1	19.9	18.3	24.9	99
WM2	63.95	6.3	21.4	101
5.5–11.0//FZ1	35.3	29.9	22.3	125
FZ2	49.75	21.5	21.7	123
11.5–27.5//HAZ1	18.9	48.7	25.4	120
HAZ2	18.0	35.5	14.2	132
>28//BM	16.5	30.1	10.3	156

Dimensionless treatment is performed for the ratio of $a_{j,I}$ and $b_{R,I}$; the closer the ratio to 1, the higher the fittingness between the two. Therefore, in Equation (4), where $1 - \frac{a_{j,i}}{b_{R,i}}$ represents the deviation of the ith characteristic quantity and $\frac{\sum_{i=1}^{n}\|1-\frac{a_{j,i}}{b_{R,i}}\|}{n}$ represents the average deviation of the sum of n characteristic quantity deviations. According to the concept of fixed moment, the fittingness was defined as $F \leq 1$. For sample 1, FZ1 acts as its statistical reference value. The fittingness between the WM and FZ of a joint is calculated using Equation (5), where n is equal to 4. The fittingness between the WM and FZ of a joint is:

$$F_{1-1} = 1 - \sqrt{\frac{\sum_{i=1}^{n}\left(1-\frac{a_{j,i}}{b_{R,i}}\right)^2}{n}} \quad (4)$$

$$= 1 - ([(1 - 19.9/35.3)^2 + (1 - 18.3/29.9)^2 + (1 - 24.9/22.3)^2 + (1 - 99/125)2]/4)^{1/2} = 0.69$$

The fittingness between the FZ and HAZ of a joint is:

$$F_{1-2} = 1 - ([(1 - 18.9/35.3)^2 + 1 - 48.7/29.9)^2 + (1 - 25.4/22.3)^2 + (1 - 120/125)^2]/4)^{1/2} = 0.60 \quad (5)$$

HAZ is derived from the evolution of the structure and performance of the BM due to a welding heat cycle. Distributions of grain sizes and hardness indicate that the structures of this zone are significantly sensitive to any change in welding conditions. As mentioned above, HAZ acts as a statistical reference for calculating the fittingness to BM based on the continuity of the evolution of the structure and performance of a joint. For HAZ1, its fittingness is calculated as:

$$F_{1-3} = 1 - ([(1 - 16.5/18.9)^2 + (1 - 30.1/48.7)^2 + (1 - 10.3/25.4)^2 + (1 - 156/120)^2]/4)^{1/2} = 0.61 \quad (6)$$

Similarly, calculation was performed for sample 2 using the same method. Based on the data (Table 4) and statistical calculations, the crystallization speed, component segregation, and other factors may lead to increased uncertainty in the recrystallization fraction, misorientation angle distribution, and other parameters of this zone during the solidification of its joint metal. There is no substantial interference from the comprehensive impact of various influencing factors and effects. For sample 1, the fittingness between WM1 and FZ1, HAZ1 and FZ1, and HAZ1 and BM were 69%, 60%, and 61%, respectively. In contrast, the fittingness between WM2 and FZ2, HAZ2 and FZ2, and HAZ2 and BM were 61%, 51%, and 81% for sample 2, respectively.

The statistical analysis results (shown in Figure 9) of the fittingness between the microstructure and performance of each zone of the joint indicate that their weakened positions can be verified using a quantitative evaluation method for fusion joints of characteristic information after the characteristic quantities of each zone are obtained. Moreover, the effects of the welding process factors can be evaluated. The mechanical properties of the samples are depicted in Figure 9b.

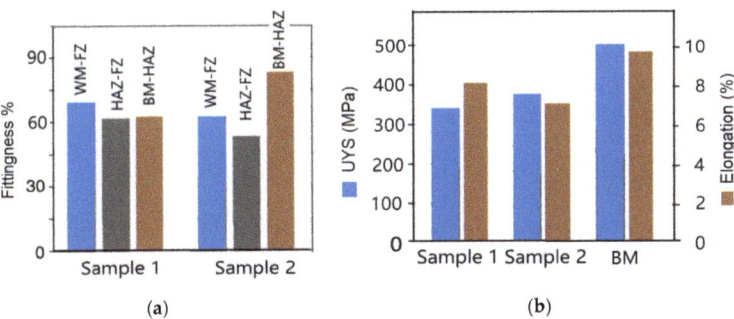

Figure 9. (a) Results of the statistical analysis of the fittingness between microstructures and performances of each zone of joints and (b) mechanical properties of samples.

As for the fusion welded joints of forged D2195 in this study, its welded joints were likely fractured under an external tensile force in the HAZ. The tensile strengths and elongations of samples 1 and 2 were 341 MPa and 8% and 370 MPa and 7%, respectively. Their strength coefficients and elongations were 69%, 75%, 84%, and 74% of those of the BM. The results were consistent with the analysis of their fittingness and prove that the comprehensive performances of forged D2195 joints were superior to those of joints of forged AA2195 fabricated using fusion metallurgy [5].

4. Conclusions

(1) The tensile strength and elongation of forged alloy D2195 joints using GTAW can be up to approximately 70% of those of the BMs fractured in the HAZs. The correlation between the joint microstructure and its performance and the weldability advantages of this type of Al alloy can be more comprehensively reflected by the recrystallization grain fraction, misorientation angle distribution, and the distribution of the grain size and Vickers hardness.

(2) Based on the concept of the Hall–Petch relationship, a statistical analysis model is established using the correlation between the microstructure and performance of several characteristic quantities by integrating the characteristic information of multiple different attributes to deeply understand the comprehensive effects of various zones of welded joints that undergo welding thermal cycles. Thus, the performance of fusion joints under the strong coupling of multiple factors can be quantitatively evaluated.

(3) The statistical calculation results of structures and fittingness between various zones of the forged D2195 joint using GTAW indicate that their weakened positions can be evaluated through a quantitative evaluation method. Thus, a cognitive approach can be adopted to not only optimize joint structures and performances by adjusting the welding processes but also popularize the application advantages of metallurgical characteristics of alloy D2195.

Author Contributions: Conceptualization, C.L. and Y.S.; methodology, formal analysis and writing—review and editing, Y.S.; validation, C.L. and Y.W.; investigation, supervision, and funding acquisition, H.L.; resources and writing—original draft preparation, C.L.; data curation and visualization, L.Y. All authors have read and agreed to the published version of the manuscript.

Funding: The authors gratefully acknowledge the financial support provided by the National Natural Science Foundation of China (Grant No. 51675375), and the project funds support provided by Science and Technology Committee of Sichuan Aerospace Technology Research Institute (Project No. F70319).

Conflicts of Interest: The authors declare no conflict of interest.

References

1. Rioja, R.J.; Liu, J. The evolution of Al–Li base products for aerospace and space applications. *Metall. Mater. Trans. A* **2012**, *A43*, 3325–3337. [CrossRef]
2. Ishchenko, A.Y. High-strength aluminium alloys for welded structures in the aircraft industry. *Weld. Int.* **2005**, *19*, 173–185. [CrossRef]
3. Prasad, N.E.; Gokhale, A.A.; Wahhil, R.J.H. *Aluminum-Lithium Alloys: Processing, Properties, and Applications*; Butterworth-Heinemann: Oxford, UK, 2014; pp. 3–22.
4. Nayan, N.; Narayana Murty, S.V.S.; Mukhopadhyay, A.K.; Prasad, K.S.; Jha, A.K.; Pant, B.; Sharma, S.C.; George, K.M. Ambient and cryogenic tensile properties of AA2195-T87 sheets with pre-aging cold work by a combination of cold rolling and stretching. *Mater. Sci. Eng. A* **2013**, *585*, 475–479. [CrossRef]
5. Abd Elaty, A.; Xu, Y.; Guo, X.; Zhang, S.-H.; Ma, Y.; Chen, D. Strengthening mechanisms, deformation behavior, and anisotropic mechanical properties of Al-Li alloys: A review. *J. Adv. Res.* **2018**, *10*, 49–67. [CrossRef] [PubMed]
6. Huang, C.; Li, H.; Li, J.; Luo, C.; Ni, Y. Residual stress measurement on propellant tank of 2219 aluminum alloy and study on its weak spot. *J. Mech. Sci. Technol.* **2017**, *31*, 2213–2220. [CrossRef]
7. Xu, Q.; Lavernia, E.J. Fundamentals of the Spray Forming Process. In Proceedings of the International Conference on Spray Deposition and Melt Atomization, SDMA 2000, Bremen, Germany, 26–28 June 2000; Volume 1, pp. 17–36.

8. Zhang, Q.; Zhang, C.; Lin, J.; Zhao, G.; Chen, L.; Zhang, H. Microstructure analysis and low-cycle fatigue behavior of spray-formed Al-Li alloy 2195 extruded plate. *Mater. Sci. Eng. A* **2019**, *742*, 773–787. [CrossRef]
9. Wang, X.D.; Pan, Q.L.; Xiong, S.W.; Liu, L.L. Prediction on hot deformation behavior of spray formed ultra-high strength aluminum alloy-A comparative study using constitutive models. *J. Alloys. Compd.* **2018**, *735*, 1931–1942. [CrossRef]
10. Singer, A.R.E. Recent developments in the spray forming of metals. *Int. J. Powder Metal. Powder Technol.* **1985**, *21*, 219–222.
11. Cordero, Z.C.; Knight, B.E.; Schuh, C.A. Six decades of the Hall-Petch effect-A survey of grain-size strengthening studies on pure metals. *Int. Mater. Rev.* **2016**, *61*, 495–507. [CrossRef]
12. Wyrzykowski, J.W.; Grabski, M.W. The Hall-Petch relation in aluminium and its dependence on the grain boundary structure. *Philos. Mag. A* **1986**, *53*, 505–520. [CrossRef]
13. Ito, Y.; Edalati, K.; Horita, Z. High-pressure torsion of aluminum with ultrahigh purity (99.9999%) and occurrence of inverse Hall-Petch relationship. *Mater. Sci. Eng. A* **2017**, *679*, 428–434. [CrossRef]
14. Xu, W.; Dvila, L.P. Tensile nanomechanics and the Hall-Petch effect in nanocrystalline aluminium. *Mater. Sci. Eng. A* **2018**, *710*, 413–418. [CrossRef]
15. Hirata, T.; Oguri, T.; Hagino, H.; Tanaka, T.; Chung, S.W.; Takigawa, Y.; Higashi, K. Influence of friction stir welding parameters on grain size and formability in 5083 aluminum alloy. *Mater. Sci. Eng. A* **2007**, *456*, 344–349. [CrossRef]
16. Naib, S.; De Waele, W.; Štefane, P.; Gubeljak, N.; Hertelé, S. Crack driving force prediction in heterogeneous welds using Vickers hardness maps and hardness transfer functions. *Eng. Fract. Mech.* **2018**, *201*, 322–335. [CrossRef]

© 2020 by the authors. Licensee MDPI, Basel, Switzerland. This article is an open access article distributed under the terms and conditions of the Creative Commons Attribution (CC BY) license (http://creativecommons.org/licenses/by/4.0/).

Article

Low Temperature Cu/Ga Solid–Liquid Inter-Diffusion Bonding Used for Interfacial Heat Transfer in High-Power Devices

Guoqian Mu, Wenqing Qu *, Haiyun Zhu, Hongshou Zhuang and Yanhua Zhang *

School of Mechanical Engineering and Automation, Beihang University, Beijing 100191, China; mgqapple@buaa.edu.cn (G.M.); zhuhaiyun@buaa.edu.cn (H.Z.); hzhuang@buaa.edu.cn (H.Z.)
* Correspondence: quwenqing@buaa.edu.cn (W.Q.); zhangyh@buaa.edu.cn (Y.Z.); Tel.: +86-10-82317702 (W.Q.)

Received: 29 July 2020; Accepted: 1 September 2020; Published: 10 September 2020

Abstract: Interfacial heat transfer is essential for the development of high-power devices with high heat flux. The metallurgical bonding of Cu substrates is successfully realized by using a self-made interlayer at 10 °C, without any flux, by Cu/Ga solid-liquid inter-diffusion bonding (SLID), which can be used for the joining of heat sinks and power devices. The microstructure and properties of the joints were investigated, and the mechanism of Cu/Ga SLID bonding was discussed. The results show that the average shear strength of the joints is 7.9 MPa, the heat-resistant temperature is 200 °C, and the thermal contact conductance is 83,541 W/(m^2·K) with a holding time of 30 h at the bonding temperature of 100 °C. The fracture occurs on one side of the copper wire mesh which is caused by the residual gallium. The microstructure is mainly composed of uniform θ-CuGa$_2$ phase, in addition to a small amount of residual copper, residual gallium and γ$_3$-Cu$_9$Ga$_4$ phase. The interaction product of Cu and Ga is mainly θ-CuGa$_2$ phase, with only a small amount of γ$_3$-Cu$_9$Ga$_4$ phase occurring at the temperature of 100 °C for 20 h. The process of Cu/Ga SLID bonding can be divided into three stages as follows: the pressurization stage, the reaction diffusion stage and the isothermal solidification stage. This technology can meet our requirements of low temperature bonding, high reliability service and interfacial heat transfer enhancement.

Keywords: heat dissipation; interfacial heat transfer; SLID; Cu/Ga; low temperature bonding; thermal interface material (TIM)

---

1. Introduction

With the increasing power dissipation and shrinking feature sizes of high-power devices (such as insulated gate bipolar translator, central processing unit, laser load devices, etc.), the heat generated is gradually increasing. High temperature has a bad impact on the performance of power devices. Research shows that the device failure rate doubles and the lifespan of the devices is halved for every 10 °C rise in the joining temperature [1], and that more than 55% of the failures of electronic devices are caused by too-high temperatures [2]. Therefore, high requirements for heat dissipation are established [3], and some efficient heat sinks (such as heat pipe, microchannel, refrigeration chip, etc.) have been developed. As such, the interfacial heat transfer between heat source and heat sink is becoming a severe bottleneck, currently limiting the further scaling of performance. Meanwhile, the heat flux density through the interface is constantly increasing [4], so interfacial heat transfer enhancement is essential for the development of high-power devices. The common methods used currently are as follows: adding thermally conductive particles (i.e., metals, ceramics) to nonmetallic thermal interface materials (TIM) [5] and using low melting temperature alloy (LMTA) [5], etc. However, the thermal conductivity of nonmetallic materials is lower than that of metallic materials, and thermal greases are easy to age. The LMTA and some nonmetallic TIM are in the liquid state in the service process, and are easy to flow out, which affects the reliability negatively [6], and external pressure is needed for them [5,6]. Thus, the metallurgical bonding of mating surfaces is the best choice for interfacial heat transfer enhancement.

Considering the special working conditions of some power devices, they must be assembled at low temperatures. For example, the aluminum ammonia heat pipe must be assembled under 95 °C, and the laser generator cannot exceed 80 °C. At the same time, they often work at a temperature that is close to or even higher than the bonding temperature. Therefore, the reliability of the bonded surfaces will not decrease at higher temperatures. At present, the methods that can be used for low-temperature bonding with high reliability include nanoparticle sintering technology, transient liquid phase sintering (TLP) and solid-liquid inter-diffusion bonding (SLID), etc.

Nanoparticle sintering technology can meet the above requirements, and the electrical and thermal conductivity [7,8] and mechanical properties [9–11] of the joints are excellent. However, its bonding temperature is generally higher than 200 °C [12], which is still too high for many power chips or devices. The interfacial microstructure has significant porosity, which will reduce the thermal and mechanical properties of the joints [13]. Besides, this technology costs a lot, given the high price of Ag. Its microstructure and properties cannot be improved under the action of high service temperature, and they will deteriorate at high temperatures and with long-term service.

Therefore, metallurgical bonding with a uniform interfacial microstructure is an important solution to this problem. The TLP sintering used mostly in the microelectronic packaging industry, also known as SLID [14], is a good way to solve this problem. It uses a high melting point metal and a low melting point metal to form intermetallic compounds (IMC), which melt at higher temperatures than the bonding temperature during isothermal solidification [15].

In this paper, the low temperature metallurgical bonding of copper substrates is investigated. The Cu substrates were successfully bonded with a self-made interlayer at 100 °C without any flux. Then, the microstructure, shear strength, temperature resistance and thermal contact conductance of the joints were studied, and the mechanism of Cu/Ga SLID bonding was discussed.

2. Materials and Methods

In this paper, we chose Ga as the low melting point metal, for the melting point of Ga (29.75 °C) is lower than Sn (231.9 °C) and In (156.6 °C), and the processing temperature of Ga-based solder is lower. Besides, gallium has excellent wetting properties and can decrease melting temperature, and so on. [16,17]. Although the eutectic points of Ga-Sn and Ga-In are lower than the melting point of Ga, the bonding temperature of Ga-based solder can completely meet our requirements. Besides, more work needs to be done on the interaction between Ga-based solder and components at low temperatures to fully utilize this kind of interesting material, and expedite its use in industry [17].

Cu substrates are utilized via their widespread use as metallization materials [18]. We chose Cu as the high melting point metal. The Au/Ga system has been investigated [19], as has the very high diffusion speed of gold into liquid Ga during the process expediting the extreme formation of kirkendall voids. The solderability of the Ag/Ga system is poor. The inter-diffusion coefficient of Cu/Ga is appropriate [20], but the Cu/Ga system has been rarely used, and they often introduce Au or Pt as a seed layer [21,22].

Cu/Ga solder paste and copper wire mesh attached with liquid Ga were used to prepare the self-made interlayer. Copper wire mesh plays the role of skeleton, which can help keep the thickness of the interlayer uniform. The Cu content in the interlayer was optimized to be about 33 wt. %, corresponding to the stoichiometry of the main reaction of copper with gallium at a small copper excess. In the whole bonding process, gallium is in the liquid state, and solid copper is surrounded by a liquid, which greatly shortens the atom diffusion distance and makes Cu/Ga inter-diffusion easy. The requirement of the substrate roughness and flatness is not high.

The copper wire mesh is pure copper, with a purity of 99.9%, 400 mesh and diameter of 40 μm. The particle size of the copper powder is 300 mesh. The purity of gallium is 99.9%. The size of the copper substrate is 60 mm × 10 mm × 1.5 mm. The materials were provided by the China General Research Institute of Nonferrous Metals. The copper substrate and copper wire mesh will be pretreated to remove the oil and oxide film on surface before joining.

Bonding process: (1) Coat the copper wire mesh with liquid Ga uniformly at 40 °C; (2) Prepare the Cu/Ga solder paste at 40 °C. Ga exists in a liquid state at 40 °C and has excellent wettability, so copper powder can be wetted by liquid Ga completely without any flux to obtain silver white Cu/Ga solder paste; (3) Apply Cu/Ga solder paste on the surface of the Cu substrates uniformly; (4) Place the copper wire mesh attached with liquid Ga between the two Cu substrates in fixture with pressure of 2 MPa; (5) Subsequently, the assembled samples are held in the furnace at 100 °C for 10 h, 20 h and 30 h respectively; (6) Take out the samples from the furnace for further observation and testing. Figure 1 shows the schematic diagram of the bonding process.

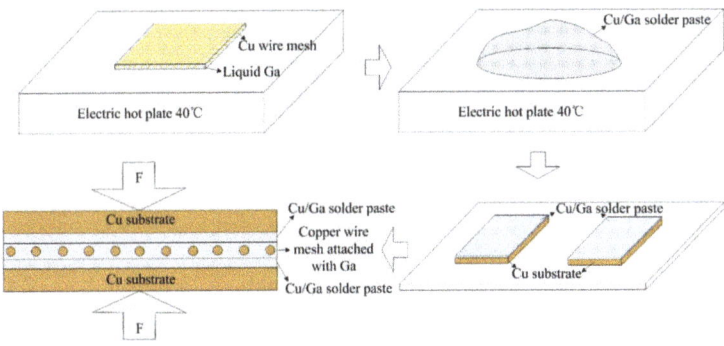

Figure 1. The schematic diagram of the bonding process.

An optical microscope (OM) (Olympus B×5) and a scanning electron microscope (SEM) (JSM-7500) were used to investigate the interfacial microstructure of the joints. The phase constitutions of the intermediate layer were analyzed by X-ray diffraction (XRD) tests with with X-ray diffractometer (D/Max-2200pc). The X-ray source is excited by the copper target. The working voltage and current were 40 kV and 40 mA, respectively, and the scanning speed was set at 6 degrees per minute. Micro-area composition analysis was investigated via Energy dispersive spectrometry (EDS) detected with SEM. The EDS analysis was performed with an acceleration voltage of 20 kV. Firstly, the joints were cut and fixed with bakelite power to make metallographic specimens. The specimens were ground with 260, 800, 1200 and 2000 waterproof abrasive paper, and then polished with 2 μm and 0.5 μm diamond paste. The XRD and SEM tests were conducted first, and etching was performed before the OM tests.

The mechanical properties of the joints were evaluated by shear testing at room temperature using a tensile testing machine (MTS50KN) with a machine displacement rate of 0.5 mm/min. The average shear strength of five joints was used to contrast the mechanical property. The joint used was the lap joint, and the lap length was 5 mm. If the tension center and the sample center are not in a straight line, the test results will be inaccurate. Two small pieces were bonded at both ends of the lap joint, as shown in Figure 2. When holding the samples on the fixture of the testing machine, attention should be paid to the position of the samples so as to prevent eccentricity.

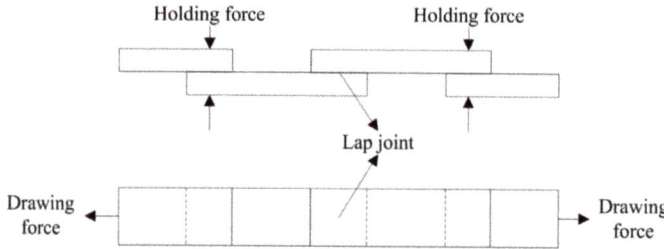

Figure 2. Schematic diagram of shear test specimen.

The method of the temperature resistance test is the weight method. A certain load is applied to the specimen, and then the specimen is heated to a predetermined temperature to observe whether the joint is remelted at this temperature. The test was carried out in a resistance furnace, and holes were punched on each piece of the lap joint respectively to hang the weight of 50 g, as shown in Figure 3. After several minutes, the specimens reached the preset temperature of 200 °C. The specimens were hold at this temperature for 10 min to observe the joints.

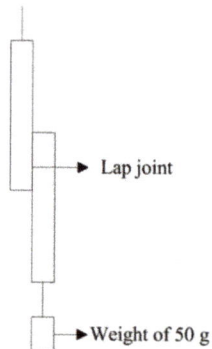

Figure 3. Schematic diagram of the temperature resistance test.

The thermal property is evaluated by the thermal contact conductance of the bonded interface, using infrared-assisted steady state temperature measurement of the vacuum thermocouple. The rated heat flux (q) was selected as 10, 50, 100, 200 and 300 W/cm^2, and the simulation device for the test was designed according to the basic law of heat conduction (Fourier law), as shown in Figure 4. The material was copper, and the thermal conductivity of Cu is 427 W/(m·K) [6]. The whole device was wrapped with multi-layer thermal insulation material and put into the vacuum tank on a special test bench for testing. The heat transfer from the hot end to the cold end can be regarded as one-dimensional axial heat conduction. From the three temperature measurement points of the cold end, the temperature value of the cold end at the bonded interface can be obtained by linear fitting. Similarly, the temperature value of the hot end at the bonded interface can be obtained. The difference between the two temperature values is the heat transfer temperature difference of the bonded interface ΔT. $q/\Delta T$ is the thermal contact conductance of the bonded interface.

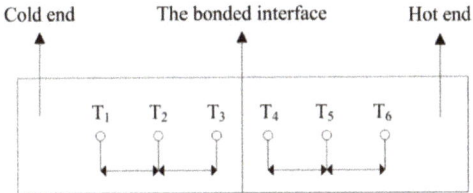

Figure 4. Schematic diagram of the simulation device for thermal property test.

3. Results

3.1. Microstructure of the Joints

The optical microscope image of the joint after bonding for 20 h at 100 °C is presented in Figure 5a. We can see that the thickness of the intermediate layer is uniform at about 120 μm. The intermediate layer can be divided into three parts: the bright yellow circular area is region A, the annular light gray phase around region A is region B, and region C is the near base metal outside region B. There is almost no impurity in region B. There are small striped or spotted dark black phases (≤5 μm) and small yellow phases in region C.

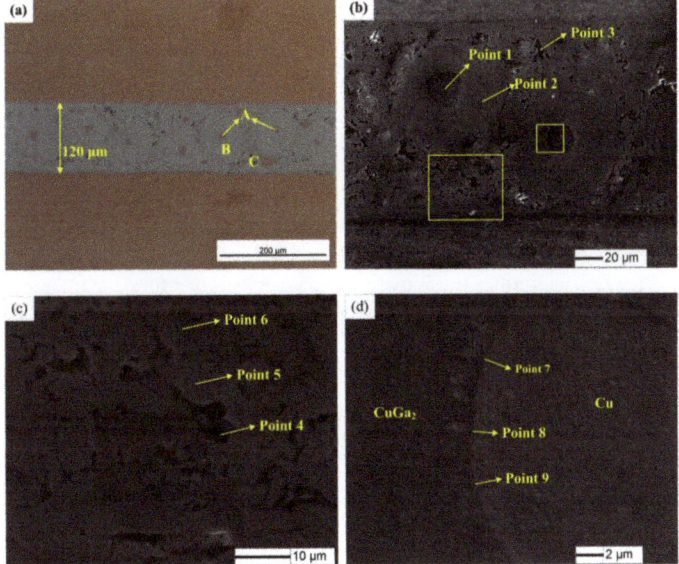

Figure 5. (a) Optical microscope image of the joint after bonding for 20 h at 100 °C; (b–d) SEM images of the joints after bonding for 20 h at 100 °C.

Figure 5b shows the SEM image of the joint after bonding for 20 h at 100 °C. Figure 5c is the magnified picture of the large rectangle in Figure 5b, and Figure 5d is the magnified picture of the small rectangle in Figure 5b. According to the EDS results as shown in Table 1, the yellow circular phase (region A) is unreacted copper wire (point 1) with a diameter of 20 μm. It indicates that about $\frac{3}{4}$ of the copper wire has been consumed. The Ga/Cu atomic ratio of the annular light gray phase (region B) around the copper wire is close to 2:1 (point 2), so may be θ-$CuGa_2$ phase, based on the Cu-Ga phase diagram [23,24]. The average Ga/Cu atomic ratio of the striped or spotted dark black phase in region C is 2.415 (point 3 and point 4), and we speculate that it may be θ-$CuGa_2$ phase with a small amount of

Ga. As its Ga/Cu atomic ratio is higher than 2, this area may be an accumulation of residual gallium. However, the EDS results show that the content of Ga here is not 100%, which may be caused by the loss of exposed Ga in the polishing process of specimens, given the low melting point and low hardness of Ga. In the intermediate layer, residual Ga is a kind of defect, which is unfavorable to the strength of the joints. Combined with the EDS results, the light gray phase (point 5) in region C should be θ-$CuGa_2$ phase, and the small yellow phase (point 6) in region C should be copper. From Figure 5d, a thin layer of dark phase can be seen at the copper wire/θ-$CuGa_2$ layer interface, and it may be γ_3-Cu_9Ga_4 phase based on the EDS results and Cu-Ga phase diagram [23,24]. This is consistent with the research that a thin layer of γ_3-Cu_9Ga_4 is formed at the Cu/$CuGa_2$ interface [25,26].

Table 1. The EDS analysis results of different points in Figure 5.

Element	Point 1	Point 2	Point 3	Point 4	Point 5	Point 6	Point 7	Point 8	Point 9
Ga	0	66.62	70.26	71.19	67.13	0	34.31	33.02	32.05
Cu	100	33.38	29.74	28.81	32.87	100	65.69	66.98	67.95
atomic ratio of Ga/Cu	–	1.99	2.36	2.47	2.04	–	0.52	0.49	0.47

Ancharov's research shows that [27,28] the only interaction product of copper with gallium is the $CuGa_2$ at temperatures near 20 °C for more than two days, and there are no other copper–gallium intermetallic compounds. Froemel found [21,29] that the $CuGa_2$ phase was formed temporarily, then a Cu_9Ga_4 phase was eventually formed; the copper and $CuGa_2$ had been completely used up and replaced by Cu_9Ga_4 phase with treatment at 200 °C for 80 h. The interfacial reaction products of Cu and Ga are γ_3-Cu_9Ga_4 and $CuGa_2$ at 200 °C for 24 h and 148 h [30]. It has been shown that [31–33] the first phase to form is usually the one that contains a higher amount of the fast-diffusing species (i.e., the lower melting point metal). Combined with the Cu-Ga binary phase diagram [23,24], it can be inferred that θ-$CuGa_2$ phase is formed first, then the γ_3-Cu_9Ga_4 phase forms by the reaction of θ-$CuGa_2$ phase and copper with the increase in diffusion temperature or the extension of holding time.

The XRD pattern of the intermediate layer is presented in Figure 6. The results show that the main phases are $CuGa_2$ phase (the atomic ratio of Ga/Cu is 2), Cu, Ga and Cu_9Ga_4 phase (the atomic ratio of Ga/Cu is 0.44), which further verifies the above analysis results. Therefore, the interaction product of Cu and Ga is mainly θ-$CuGa_2$ phase, with a little amount of γ_3-Cu_9Ga_4 phase at the temperature of 100 °C for 20 h.

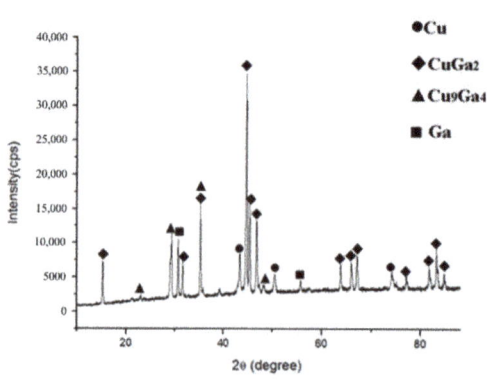

Figure 6. XRD results of the intermediate layer.

Figure 7a–c shows the optical microscope image of the joints at 100 °C with different holding times. From Figure 7a, we can see that the interdiffusion of Cu and Ga is insufficient, as large amounts of residual copper powder and residual gallium exist at the same time, and the diameter of copper wire is about 25 µm. With the prolongation of holding time, the area of region C and the quantity of residual Ga is gradually reduced, and θ-CuGa$_2$ phase (region B) increases in quantity. For 30 h (in Figure 7c), the annular θ-CuGa$_2$ phases (region B) connect with each other to form a continuous θ-CuGa$_2$ layer, and residual Ga mainly concentrates near the base metal, which may be caused by the insufficient diffusion of base metal and liquid Ga. After the annealing treatment of the Cu substrates (in Figure 7d), the residual gallium in region C is reduced significantly, which verifies our above assumption. At this time, the microstructure of the joint is mainly composed of θ-CuGa$_2$ phase, and θ-CuGa$_2$ has some advantages over other IMCs, including the nearly isotropic thermal expansion and mechanical properties, small Young's modulus and hardness, and increased compliance and softness [34]. Compared with Figure 7b,c, the copper wire diameter changes a little, which indicates that the diffusion reaction becomes slow after a certain time. The reaction layer thickness grows in proportion to the square root of the reaction time [30], and thus grows slow later. As a whole, with the prolongation of the holding time, the number of defects in the joint decreases and the microstructure becomes more uniform.

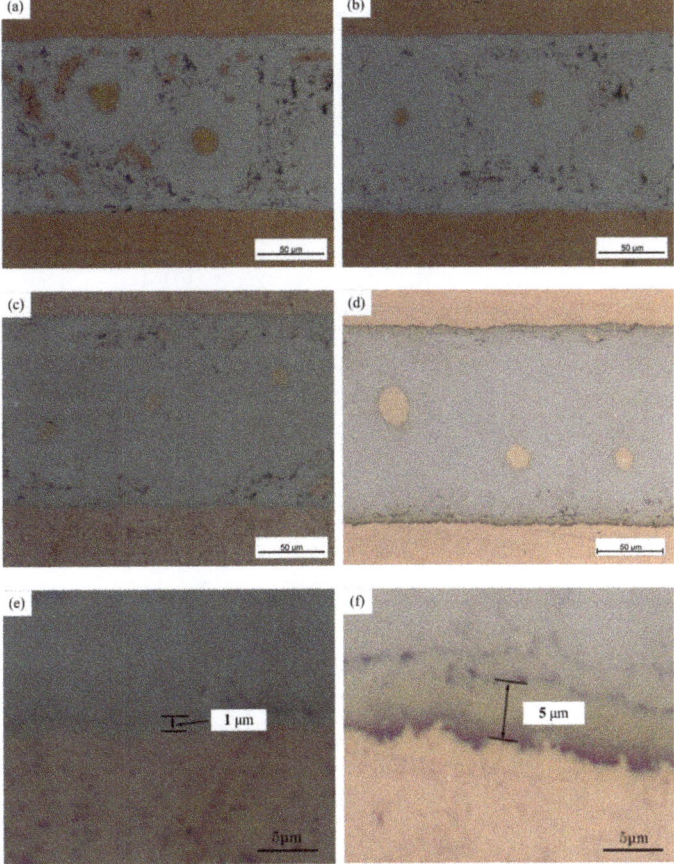

Figure 7. Optical microscope image of the joints at 100 °C. (**a**) Holding time of 10 h; (**b**) Holding time of 20 h; (**c**) Holding time of 30 h; (**d**) After annealing treatment of Cu substrates with holding time of 30 h; (**e**) Interface between copper substrate and the intermediate layer in Figure 4c; (**f**) Interface between copper substrate and the intermediate layer in Figure 4d.

Figure 7e,f shows the interface between copper substrate and the intermediate layer with different treatments of the base metal. Without the annealing treatment of Cu substrates, the thickness of the interface is only about 1 µm, and the interface is relatively flat. After annealing, the thickness of the interface is about 5 µm, and the interface is jagged as a mountain peak. The annealing treatment is helpful for the interdiffusion of gallium and the base metal, which is favorable for reducing the defects caused by residual gallium in the intermediate layer.

3.2. Properties of the Joints

Figure 8 shows the variation of shear strength and thermal contact conductance with holding time. The average shear strength and the thermal contact conductance increase gradually with the extension of the holding time. The maximum average shear strength of the joint is 7.9 MPa, and the thermal contact conductance is 88,315 W/(m^2·K) with a holding time of 30 h at a bonding temperature of 100 °C. The fracture occurs on one side of the copper wire mesh along line A or line B in Figure 9, which is caused by the regular distribution of residual Ga. Residual Ga mainly distributes in region C near the base metal, and uniform θ-CuGa$_2$ phase is formed around the copper wire mesh, so region C is the weakest area of the joint. Reducing the size and quantity of residual Ga in the intermediate layer is an effective way to improve the strength of the joints.

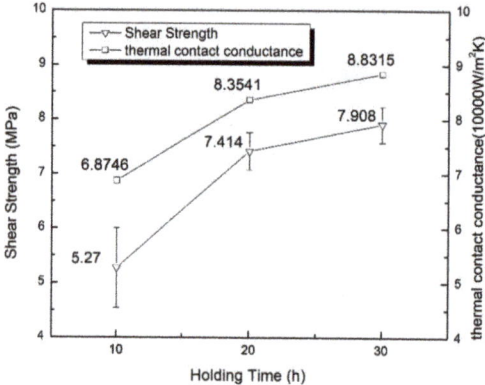

Figure 8. Effect of holding time on shear strength and thermal contact conductance of the joints at the bonding temperature of 100 °C.

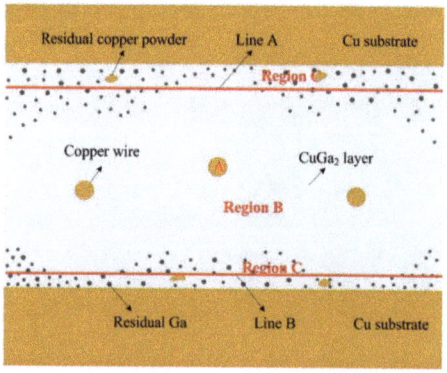

Figure 9. Diagram of fracture location in shear strength test.

The results of the temperature resistance test show that there is no remelting or liquefying phenomenon in the joints with the holding times of 20 h and 30 h, which can meet our requirements. In fact, as the intermediate layer is mainly composed of θ-CuGa$_2$ phase with a melting point of 254 °C, the joint can surely withstand the temperature of 200 °C when there is only a little residual Ga.

With the prolongation of holding time, the strength and the thermal contact conductance of the joints were improved, which indicated that the thickening of the θ-CuGa$_2$ layer (region B) and the decrease of residual gallium (region C) were beneficial for the mechanical and thermal properties of the joints. This is because region B is a single θ-CuGa$_2$ phase with a homogeneous and dense microstructure; region C contains θ-CuGa$_2$ phase, residual copper powder and residual gallium without a homogeneous and dense microstructure, and there are many interfaces between the different phases. Besides this, the mechanical property of gallium is low. Therefore, the reliability of the joints will not decrease for a long time in service.

4. Discussion

During the heat preservation process, liquid gallium and solid copper wire, and copper powder and copper substrate, interdiffuse and react with each other. The whole process of Cu/Ga SLID bonding is divided into the following three stages:

(1) Pressurization stage—Apply a certain amount of pressure to the joint in the heat preservation process. Liquid Ga is forced to flow into the gap in the copper wire mesh under the pressure to discharge the air mixed in the bonding process, thus the pressure in the interlayer tends to be the same. It can also increase the pressure of the solid/liquid interface, so that the contact between the solid copper and liquid gallium is more sufficient for subsequent diffusion;

(2) Reaction diffusion stage—Driven by a high temperature and concentration gradient, surface diffusion occurred rapidly at the solid Cu/liquid Ga interface to form θ-CuGa$_2$ phase. The θ-CuGa$_2$ phase formed due to the fast diffusion kinetics of Cu to liquid gallium [35]. The copper atoms on the solid copper surface diffuse into the surrounding liquid Ga and react rapidly with it to form θ-CuGa$_2$, as shown in Figure 10a. With the increase in the θ-CuGa$_2$ compound, θ-CuGa$_2$ crystals begin to precipitate, as shown in Figure 10b. Then, the θ-CuGa$_2$ layer gradually forms, as shown in Figure 10c. Above is the formation process of intermetallic compound θ-CuGa$_2$. After that, the solid copper and liquid gallium are separated by a θ-CuGa$_2$ layer, and the reaction diffusion will be hindered, as shown in Figure 10d. At this time, the θ-CuGa$_2$ phase may contain a higher amount of Ga [35].

Froemel thinks [21] that the increased diffusion of the solid phase into the liquid phase leads to a rapid increase of the Cu atoms inside the liquid Ga. After reaching a saturation of solubility, intermetallic compounds are formed, and thus solidification happens. According to the Cu/Ga binary phase diagram [23,24], when the concentration of copper is more than 1%, the liquid's temperature will exceed 100 °C. This means that there will be a solid phase when the concentration of copper is 1~33.3% at 100 °C if the Cu/Ga solution is formed first, which is contrary to the fact. Therefore, in the early stage of the diffusion, the copper atoms react immediately with Ga to form θ-CuGa$_2$ phase, and do not form the Cu/Ga solution;

(3) Isothermal solidification stage—This stage takes a long time. The main mode of the atoms passing through the θ-CuGa$_2$ layer is grain boundary diffusion and vacancy diffusion. In the early stage of the θ-CuGa$_2$ layer formation, the grains grow up gradually, the grain boundary gap is large, most of the grain boundaries are large angle grain boundaries, and the grain boundary energy is high. When the concentration gradient is certain, the activation energy required for copper atoms to cross the grain boundaries is low, which makes it easy to diffuse through the grain boundaries. At the same time, gallium atoms also move across the grain boundaries to the solid copper. At high temperatures, there is a certain concentration of vacancy in the θ-CuGa$_2$ crystal. Driven by the concentration gradient, copper atoms can also migrate to liquid gallium through vacancy diffusion. In later stages of grain boundary diffusion, vacancy diffusion will play a more important role.

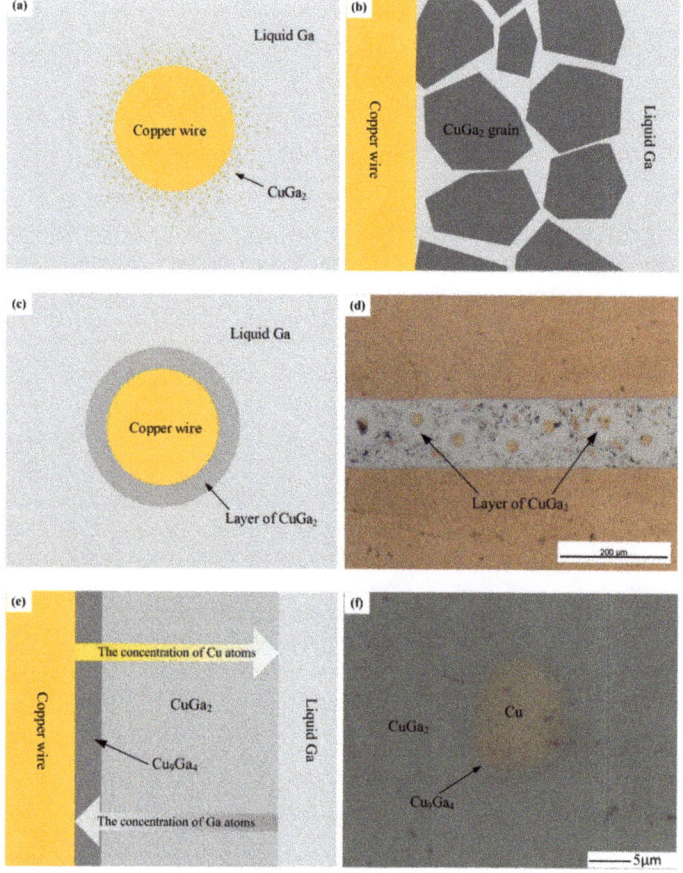

Figure 10. Schematic of the mechanism of Cu/Ga SLID bonding. (**a–c**) Formation process of intermetallic compound θ-CuGa$_2$; (**d**) Formation of θ-CuGa$_2$ layer; (**e**) Diffusion of elements in isothermal solidification stage; (**f**) Formation of thin γ$_3$-Cu$_9$Ga$_4$ layer.

Copper atoms migrate from the solid copper surface, across the θ-CuGa$_2$ layer, to liquid gallium by grain boundary diffusion and vacancy diffusion. As shown in Figure 10e, from left to right on the θ-CuGa$_2$ layer, the concentration of copper atoms decreases gradually, and the concentration of gallium atoms increases gradually. Because the copper atom content in the θ-CuGa$_2$ layer on the side of the copper wire is relatively high, a new intermetallic compound, γ$_3$-Cu$_9$Ga$_4$ (2θ-CuGa$_2$+7Cu = γ$_3$-Cu$_9$Ga$_4$), is gradually generated. The γ$_3$-Cu$_9$Ga$_4$ generated is located at the interface of the copper wire and θ-CuGa$_2$ layer. Then, a thin γ$_3$-Cu$_9$Ga$_4$ layer is gradually formed as time goes on, as shown in Figure 10f. As the gallium atom content in the θ-CuGa$_2$ layer on the side of liquid gallium is high, Ga reacts with the diffused copper atom to form θ-CuGa$_2$. On the micro level, θ-CuGa$_2$ grains grow and the θ-CuGa$_2$ grain boundaries become more flat and narrower. On the macro level, the solid–liquid interface advances in the direction of liquid gallium. Ideally, all liquid Ga in the interlayer will be completely transformed into the solid phase if the holding time is long enough.

5. Conclusions

In this work, Cu substrates were successfully bonded using a self-made interlayer at 100 °C without any flux by Cu/Ga SLID bonding. The microstructure, shear strength, temperature resistance and thermal contact conductance of the joints were studied, and the mechanism of Cu/Ga SLID bonding was discussed. This self-made interlayer can be used as a kind of thermal interface material to facilitate interfacial heat transfer. This low temperature metallurgical bonding technology can provide some reference for solving the problem of the heat dissipation of some high-power devices with high reliability. The following conclusions can be drawn:

1. The interfacial microstructure is θ-$CuGa_2$ phase, residual copper, residual gallium and γ_3-Cu_9Ga_4 phase. A continuous θ-$CuGa_2$ layer is formed around the copper wire, which contributes to the main interfacial microstructure. The residual gallium is mainly distributed near the base metal, far away from the copper wire, which is caused by the insufficient interdiffusion of copper substrate and gallium. A thin γ_3-Cu_9Ga_4 layer is formed at the copper wire/θ-$CuGa_2$ layer interface.
2. The interaction product of Cu and Ga is mainly θ-$CuGa_2$ phase, with only a small amount of γ_3-Cu_9Ga_4 phase occurring at the temperature of 100 °C for 20 h. With the prolongation of holding time, the θ-$CuGa_2$ layer thickens, residual Ga decrease, and the microstructure is more uniform.
3. The low-temperature metallurgical bonding of Cu substrates is successfully realized using a self-made interlayer at 100 °C without any flux. The maximum shear strength is 7.9 MPa, the heat-resistant temperature is 200 °C, and the thermal contact conductance is 88,315 W/(m^2·K) with a holding time of 30 h at a bonding temperature of 100 °C. The fracture occurs on one side of the copper wire mesh, which is caused by the regular distribution of residual Ga. Region C is the weakest area of the joint. With the prolongation of holding time, the properties of the joints are improved. The thickening of the θ-$CuGa_2$ layer and reduction in residual gallium are beneficial for the mechanical and thermal properties of the joints.
4. The process of the Cu/Ga SLID bonding can be divided into three stages: pressurization stage, reaction diffusion stage and isothermal solidification stage. In the reaction diffusion stage, Cu reacts with Ga to form θ-$CuGa_2$ phase, and then a θ-$CuGa_2$ layer is formed gradually. In the isothermal solidification stage, Cu reacts with θ-$CuGa_2$ to form γ_3-Cu_9Ga_4 phase, and a γ_3-Cu_9Ga_4 layer is gradually formed at the copper wire/θ-$CuGa_2$ layer interface; the diffused Cu atoms react with liquid Ga to form θ-$CuGa_2$, which thickens the θ-$CuGa_2$ layer on the liquid Ga side.

Author Contributions: Conceptualization, W.Q. and H.Z. (Hongshou Zhuang); methodology, W.Q. and G.M.; validation, W.Q. and Y.Z.; formal analysis, W.Q. and G.M.; investigation, G.M. and H.Z. (Haiyun Zhu); resources, G.M. and H.Z. (Haiyun Zhu); data curation, G.M.; writing—original draft preparation, G.M.; writing—review and editing, W.Q. and Y.Z.; supervision, W.Q. and Y.Z.; project administration, W.Q.; funding acquisition, W.Q. All authors have read and agreed to the published version of the manuscript.

Funding: This project was funded by key special projects of the national key R&D program (2017YFB0305700).

Acknowledgments: All the authors of this paper sincerely thank the teachers of the testing center of the school of materials science and engineering in Beihang university, and the China academy of space technology for their help in performance testing. This work was supported by key special projects of the national key R&D program (2017YFB0305700).

Conflicts of Interest: The authors declare no conflict of interest.

References

1. Belhardja, S.; Mimounia, S.; Saidanea, A.; Benzohra, M. Using microchannels to cool microprocessors: A transmission-line-matrix study. *Microelectron. J.* **2003**, *34*, 247–253. [CrossRef]
2. Yeh, L.T. Review of heat transfer technologies in electronic equipment. *J. Electron. Packag. Trans. ASME* **1995**, *117*, 333–339. [CrossRef]

3. Moore, A.L.; Shi, L. Emerging challenges and materials for thermal management of electronics. *Mater. Today* **2014**, *17*, 163–174. [CrossRef]
4. Ebadian, M.A.; Lin, C.X. A review of high-heat-flux heat removal technologies. *J. Heat Transf.* **2011**, *133*, 110801. [CrossRef]
5. Gwinn, J.P.; Webb, R.L. Performance and testing of thermal interface materials. *Microelectron. J.* **2003**, *34*, 215–222. [CrossRef]
6. Hansson, J.; Torbjörn, M.J.N.; Ye, L.L.; Liu, J. Novel nanostructured thermal interface materials: A review. *Int. Mater. Rev.* **2017**, *63*, 22–45. [CrossRef]
7. Mei, Y.H.; Wang, T.; Cao, X.; Chen, G.; Lu, G.Q.; Chen, X. Transient Thermal Impedance Measurements on Low-Temperature-Sintered Nanoscale Silver Joints. *J. Electron. Mater.* **2012**, *41*, 3152–3160. [CrossRef]
8. Bai, J.G.; Zhang, Z.Z.; Calata, J.N.; Lu, G. Low-Temperature Sintered Nanoscale Silver as a Novel Semiconductor Device-Metallized Substrate Interconnect Material. *IEEE Trans. Compon. Packag. Technol.* **2006**, *29*, 589–593. [CrossRef]
9. Fu, S.C.; Mei, Y.H.; Lu, G.Q.; Li, X.; Chen, G.; Chen, X. Pressureless sintering of nanosilver paste at low temperature to join large area (≥100 mm^2) power chips for electronic packaging. *Mater. Lett.* **2014**, *128*, 42–45. [CrossRef]
10. Chen, G.; Yu, L.; Mei, Y.H.; Li, X.; Chen, X.; Lu, G.Q. Uniaxial ratcheting behavior of sintered nanosilver joint for electronic packaging. *Mater. Sci. Eng. A* **2014**, *591*, 121–129. [CrossRef]
11. Yan, J.F.; Zou, G.S.; Wu, A.P.; Ren, J.L.; Yan, J.C.; Hu, A.M.; Zhou, Y. Pressureless bonding process using Ag nanoparticle paste for flexible electronics packaging. *Scr. Mater.* **2012**, *66*, 582–585. [CrossRef]
12. Wang, S.A.; Li, M.Y.; Ji, H.J.; Wang, C.Q. Rapid pressureless low-temperature sintering of Ag nanoparticles for high-power density electronic packaging. *Scr. Mater.* **2013**, *69*, 789–792. [CrossRef]
13. Youssef, T.; Rmili, W.; Woirgard, E.; Azzopardi, S.; Vivet, N.; Martineau, D.; Meuret, R.; Quilliec, G.L.; Richard, C. Power modules die attach: A comprehensive evolution of the nanosilver sintering physical properties versus its porosity. *Microelectron. Reliab.* **2015**, *55*, 1997–2002. [CrossRef]
14. Sun, L.; Chen, M.H.; Zhang, L.; He, P.; Xie, L.S. Recent progress in SLID bonding in novel 3D-IC technologies. *J. Alloys Compd.* **2020**, *818*, 152825. [CrossRef]
15. Jung, D.H.; Sharma, A.; Mayer, M.; Jung, J.P. A review on recent advances in transient liquid phase (TLP) bonding for thermoelectric power module. *Rev. Adv. Mater. Sci.* **2018**, *53*, 147–160. [CrossRef]
16. Matsushita, M.; Sasaki, Y.; Ikuta, Y. Investigation of a New Sn–Cu–Ga Alloy Solder. *Defect Diffus. Forum* **2011**, *312*, 518–523. [CrossRef]
17. Liu, S.; Sweatman, K.; McDonald, S.; Nogita, K. Ga-Based Alloys in Microelectronic Interconnects: A Review. *Materials* **2018**, *11*, 1384. [CrossRef]
18. Ramm, P.; Lu, J.J.Q.; Taklo, M.M.V. Thermocompression Cu–Cu Bonding of Blanket and Patterned Wafers. In *Handbook of Wafer Bonding*; Wiley-VCH Verlag & Co. KGaA: Weinheim, Germany, 2012; pp. 161–163.
19. Frömel, J.; Lin, Y.C.; Wiemer, M.; Gessner, T.; Esashi, M. Low temperature metal interdiffusion bonding for micro devices. In Proceedings of the 2012 3rd IEEE International Workshop on Low Temperature Bonding for 3D Integration, Tokyo, Japan, 22–23 May 2012; p. 163.
20. Marinković, Ž.; Simić, V. Comparative analysis of interdiffusion in some thin film metal couples at room temperature. *Thin Solid Film* **1992**, *217*, 26–30. [CrossRef]
21. Froemel, J.; Baum, M.; Wiemer, M.; Gessner, T. Low-temperature wafer bonding using solid-liquid inter-diffusion mechanism. *J. Microelectromech. Syst.* **2015**, *24*, 1973–1979. [CrossRef]
22. Lin, S.K.; Chang, H.M.; Cho, C.L.; Liu, Y.C.; Kuo, Y.K. Formation of solid-solution Cu-to-Cu joints using Ga solder and Pt under bump metallurgy for three-dimensional integrated circuits. *Electron. Mater. Lett.* **2015**, *11*, 687–694. [CrossRef]
23. Ma, C.; Xue, S.; Wang, B. Study on novel Ag-Cu-Zn-Sn brazing filler metal bearing Ga. *J. Alloys Compd.* **2016**, *688*, 854–862. [CrossRef]
24. Li, J.B.; Ji, L.N.; Liang, J.K.; Zhang, Y.; Luo, J.; Li, C.R.; Rao, G.H. A thermodynamic assessment of the copper–gallium system. *Calphad* **2008**, *32*, 447–453. [CrossRef]
25. Liu, S.; Zeng, G.; Yang, W.; McDonald, S.; Gu, Q.; Matsumura, S.; Nogita, K. Interfacial Reactions between Ga and Cu-10Ni Substrate at Low Temperature. *ACS Appl. Mater. Interfaces* **2020**, *12*, 21045–21056. [CrossRef] [PubMed]

26. Lin, S.K.; Cho, C.L.; Chang, H.M. Interfacial Reactions in Cu/Ga and Cu/Ga/Cu Couples. *J. Electron. Mater.* **2013**, *43*, 204–211. [CrossRef]
27. Ancharov, A.I.; Grigoryeva, T.F.; Barinova, A.P.; Boldyrev, V.V. Interaction between copper and gallium. *Russ. Metall. (Met.)* **2008**, *2008*, 475–479. [CrossRef]
28. Ancharov, A.I.; Grigoryeva, T.F. Investigation of the mechanism of interaction between reagents in alloys based on Cu-Ga system. *Nucl. Instrum. Methods Phys. Res. A* **2005**, *543*, 139–142. [CrossRef]
29. Froemel, J.; Baum, M.; Wiemer, M.; Gessner, T. Solid liquid inter-diffusion bonding at low temperature. In Proceedings of the 2014 4th IEEE International Workshop on Low Temperature Bonding for 3D Integration (LTB-3D), Tokyo, Japan, 15–16 July 2014; p. 62.
30. Chen, S.W.; Lin, J.M.; Yang, T.C.; Du, Y.H. Interfacial Reactions in the Cu/Ga/Co and Cu/Ga/Ni Samples. *J. Electron. Mater.* **2019**, *48*, 3643–3654. [CrossRef]
31. Cardellini, F.; Contini, V.; Mazzone, G.; Vittori, M. Phase Transformations and chemical reactions in mechanically alloyed Cu/Zn powders. *Scr. Metall. Mater.* **1993**, *28*, 1035–1038. [CrossRef]
32. Pabi, S.K.; Joardar, J.; Murty, B.S. Formation of nanocrystalline phases in the Cu-Zn system during mechanical alloying. *J. Mater. Sci.* **1996**, *31*, 3207–3211. [CrossRef]
33. Pretorius, R.; Theron, C.C.; Vantomme, A.; Mayer, J.W. Compound Phase Formation in Thin Film Structures. *Crit. Rev. Solid State Mater. Sci.* **1999**, *24*, 1–62. [CrossRef]
34. Liu, S.; Yang, W.; Kawami, Y.; Gu, Q.; Matsumura, S.; Qu, D.; McDonald, S.; Nogita, K. Effects of Ni and Cu Antisite Substitution on the Phase Stability of $CuGa_2$ from Liquid Ga/Cu-Ni Interfacial Reaction. *ACS Appl. Mater. Interfaces* **2019**, *11*, 32523–32532. [CrossRef] [PubMed]
35. Hong, S.J.; Suryanarayana, C. Mechanism of low-temperature θ-$CuGa_2$ phase formation in Cu-Ga alloys by mechanical alloying. *J. Appl. Phys.* **2004**, *96*, 6120–6126. [CrossRef]

© 2020 by the authors. Licensee MDPI, Basel, Switzerland. This article is an open access article distributed under the terms and conditions of the Creative Commons Attribution (CC BY) license (http://creativecommons.org/licenses/by/4.0/).

Article

Electrochemical Migration Inhibition of Tin by Disodium Hydrogen Phosphate in Water Drop Test

Bokai Liao [1,*], Hong Wang [1], Shan Wan [1], Weiping Xiao [2] and Xingpeng Guo [1,3,*]

[1] Guangzhou Key Laboratory for Clean Energy and Materials, Institute of Clean Energy and Materials, School of Chemistry and Chemical Engineering, Guangzhou University, Guangzhou 510006, China; honggwang@yeah.net (H.W.); wanshan1145@sina.com (S.W.)
[2] College of Science, Nanjing Forestry University, Nanjing 210037, China; wpxiao@njfu.edu.cn
[3] School of Chemistry and Chemical Engineering, Huazhong University of Science and Technology, Wuhan 430074, China
* Correspondence: bokailiao@gzhu.edu.cn (B.L.); guoxp@mail.hust.edu.cn (X.G.);
 Tel.: +8-613-277-977-921 (B.L.); +8-618-998-821-598 (X.G.)

Received: 12 June 2020; Accepted: 10 July 2020; Published: 14 July 2020

Abstract: The inhibition effect of Na_2HPO_4 on the electrochemical migration (ECM) of pure tin was investigated by means of water drop testing and surface characterizations. The effects of concentration of Na_2HPO_4 and applied direct current (DC) bias voltage on the ECM were also studied. Results showed that the mean time to failure caused by ECM decreased with the increasing bias voltage. Upon addition of relative high concentrations of Na_2HPO_4, Na_2HPO_4 can react with metallic tin or tin ions to form a protective film on the surface of anode and increase the pitting potential. The rate of anodic dissolution can be slowed down and thus ECM of tin was retarded. Fractal-like dendrites formed after ECM tests in the absence and presence of low concentrations of Na_2HPO_4 mainly consisted of tin elements. Relevant reactions were proposed to explain the inhibitory effect of Na_2HPO_4 on the ECM of tin.

Keywords: tin; electrochemical migration; dendrite; inhibitor

1. Introduction

With the advent of microelectronics over the past few decades, electronic components are becoming more sensitive to corrosion due to miniaturization and integration processes [1]. As for various kinds of electronic materials, such as tin, copper, nickel, silver and their alloys [2], the common failure behaviors include: atmospheric corrosion [3], electrochemical migration (ECM) [4], formation of conductive filament [5] and deformation [6], etc. ECM can be defined as a form of corrosion caused by the applied bias voltage, which can result in insulation resistance degradation or short circuit in electronic devices [7]. The required conditions for ECM mainly include bias voltage between two electrodes, elevated relative humidity, temperature and time. Three basic steps consist of ECM, including anodic dissolution, migration of metal ions and deposition of metal ions.

Due to excellent electric conductivity, low melting point, moderate corrosion and oxidation resistance, tin-based alloys were commonly utilized in practical electronic connections [8]. This has made the research of reliability for tin-based solder alloys an important issue for electronic assembly and encapsulation [9]. Numerous authors have reported on atmospheric corrosion [10], mechanism of ECM [11], and the effect of some environmental or material factors on ECM (such as the addition of alloy elements [12], type of the applied bias voltage [13], thickness of the absorbed electrolyte [14], typical contaminants [15], etc.). Tin and tin-based alloys are sensitive to ECM [16], but few methods for mitigating ECM of tin and tin-based alloys have been established [17].

Phosphate compounds, including disodium hydrogen phosphate (Na_2HPO_4) [18], zinc phosphate [19], cerium dibutylphosphate [20], lithium zinc phosphate [21], etc., were widely used as environmentally-friendly inorganic inhibitors due to their low toxicity and low cost [22]. Awad et al. [23] found that tertiary phosphate ions exhibited a corrosion inhibition effect on tin owing to the formation of a passive film on the surface of electrode. Yohai et al. investigated the competitive adsorption inhibition mechanism of phosphate by changing $[PO_4^{3-}]/[Cl^-]$ and $[Cl^-]/[OH^-]$ ratios. It was found that tin phosphate complexes can retard the anodic dissolution and migration of metal ions as phosphate corroded metallic tin [24]. Na_2HPO_4 can be used as an inhibitor for tin in a proper concentration [25]. In this case, phosphate compounds can be used as an alternative inhibitor for ECM of tin by forming a protective film on the anode and retarding the migration behavior of tin ions.

The objective of the present work is to evaluate the inhibition effect of Na_2HPO_4 on the ECM of tin. The effects of Na_2HPO_4 concentration and applied DC bias voltage on the ECM of tin were studied. Potentiodynamic polarization was used to investigate the effect of Na_2HPO_4 on the anodic and cathodic reactions. The surface characterizations of scanning electron microscopy (SEM) and energy-dispersive spectrometry (EDS) were used to clarify the relevant mechanisms. This study can provide a basis for the development of phosphate compound-based ECM inhibitors for tin and tin alloys.

2. Materials and Methods

2.1. Materials and Solution Preparation

The tin samples were processed from commercial pure Sn (>99.999%, mass%) (Sichuan Xinlong Hoof Technology Development Co. Ltd., Sichuan, China) with dimensions of 2 mm × 5 mm × 10 mm. Each electrode for ECM test had two tin samples (one for the anode and one for the cathode) sealed in a cylindrical plastic tube using epoxy resin. The gap size was 0.5 mm in the parallel direction and the working area was 0.1 cm². The sealed tin electrodes for polarization curve measurements had a working area of 0.5 cm². Stranded copper wires were welded to the back of each electrode to ensure electrical contact. All the test surfaces were successively polished with 400, 800, 1200 grit silicon carbide papers, then sonicated continuously in de-ionized water and acetone each time before tests.

Na_2HPO_4 (Sigma, analytical grade) and NaCl (Sigma, analytical grade) solutions were prepared with de-ionized water (Resistivity of 18.2 MΩ cm). A DDS-307A conductivity tester (Shanghai Rex Instrument Factory, China) and a PHS-3C pH meter (Shanghai Rex Instrument Factory, China) were used to record the variations in the conductivity and pH value of the solutions, respectively.

2.2. Setup of Water Drop Test With and Without Na_2HPO_4 and ECM Measurements

The water drop (WD) test has been considered as a realistic method to simulate the ECM process in the case of droplets splashing on the surface of electronics. As shown in Figure 1, the setup of the WD test mainly consists of a CS350 electrochemical workstation (Wuhan Corrtest, China) and a digital microscopy (Betical XTL-6745J4, Nanjing, China). Prior to bias, a droplet of well-defined electrolyte was placed on the surface of electrodes using a pipetting device. A direct current (DC) bias voltage (3 V, 5 V, 8 V or 10 V) was applied between the twin electrodes. The current flowing was recorded as a function of time by the electrochemical workstation. The change of morphology on the electrode surface was in situ recorded by the digital microscope. Considering the vaporization process of droplets, the longest time for ECM test was set as 3000 s. All the ECM tests were repeated at least five times to check reproducibility.

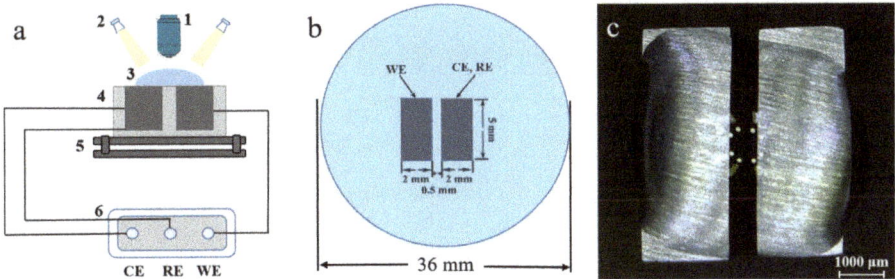

Figure 1. Schematic diagram of setup for electrochemical migration using the water drop test (WDT): (**a**) (1) digital microscope; (2) cold light sources; (3) waterdrop; (4) electrodes; (5) horizontal stage; (6) electrochemical workstation; (**b**) vertical view of electrode; (**c**) optical photo of waterdrop before electrochemical migration (ECM) test.

2.3. Polarization Curves and Cyclic Voltammetry Measurements

A three-electrode single-compartment glass cell was employed for the electrochemical measurements at ambient temperature (24 ± 0.5 °C), including a Pt foil (2 cm^2) as a counter electrode and a saturated calomel electrode (SCE) as a reference electrode. Polarization curves were performed in solutions containing various concentrations of Na_2HPO_4 with a scan rate of 0.5 mV/s from −0.15 V to 1.75 V versus the open-circuit potential.

2.4. Surface Characterization

The microstructure and composition of dendrite and precipitate obtained after ECM tests were examined by scanning electron microscopy (SEM; JSM-6701F, Hitachi Ltd., Tokyo, Japan) quipped energy-dispersive spectrometry (EDS; Oxford INCA energy 300, Oxford Instruments, Oxford, UK).

3. Results and Discussion

3.1. Effect of the Alert of Solution Chemistry on the Probability of the ECM of Sn

Figure 2 shows the change of pH value and conductivity of solutions containing various concentrations of Na_2HPO_4. As shown in Reactions (1) and (2) [26], hydrolysis of Na_2HPO_4 in water gave rise to the increase in concentration of OH$^-$ and some other ions (such as Na, $H_2PO_4^-$, etc.). The pH value and conductivity of solution increased with the increasing concentration of Na_2HPO_4. For example, as the concentration of Na_2HPO_4 increases from 0 mg/L to 1000 mg/L, pH value increases from 6.2 to 8.9 while conductivity changes from 0.07 to 142.4 µS/cm.

$$HPO_4^{2-} + H_2O \rightleftharpoons H_2PO_4^- + OH^- \tag{1}$$

$$H_2PO_4^- + H_2O \rightleftharpoons H_3PO_4 + OH^- \tag{2}$$

Normally, the rates of anodic and cathodic reactions process increase resulting from the decrease of solution resistance during ECM [27]. The rising of conductivity accelerates the growth of dendrite, but the relatively higher alkaline pH condition does not favor dendrite formation [28].

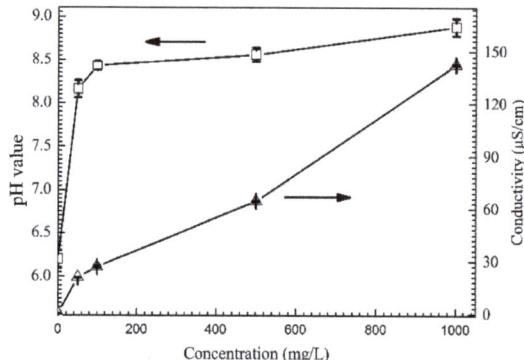

Figure 2. Change of pH value and conductivity with the concentration of Na_2HPO_4.

3.2. Effect of Na_2HPO_4 Concentration on the Probability of the ECM of Sn at Different DC Bias Voltages

Figure 3 shows the influence of Na_2HPO_4 concentration on the current density during ECM under a droplet of 15 µL at different DC bias voltages. During the ECM test, the dendrite grows from a cathode and then bridges the anode causing a short circuit [29]. The short circuit results in a sudden increase of current density up to one or two orders of magnitude. The time of short circuit can be defined as the time to failure (TTF) caused by ECM [30]. As shown in Figure 3a, in the absence of Na_2HPO_4, mean TTF of ECM for tin descended with the increase of DC bias voltage. For example, mean TTF descended from 31.9 s to 1.7 s while DC bias voltage increased from 3 V to 10 V. At low concentration levels of Na_2HPO_4 (such as 50 mg/L and 100 mg/L), results of mean TTF are similar with that obtained without Na_2HPO_4, as shown in Figure 3b and c. For example, under the 3 V bias condition, mean TTF is 32.23 s under a droplet without Na_2HPO_4 while mean TTF is 28.83 s in the presence of 50 mg/L and mean TTF is 33.02 s in the presence of 100 mg/L. At high concentration levels of Na_2HPO_4 (such as 500 mg/L and 1000 mg/L), the sudden increase of current density does not occur after 3000 s at different DC bias voltages, indicating that Na_2HPO_4 can inhibit the growth of the dendrite during ECM tests. As shown in Figure 3d,e, the significant fluctuation of current density can mainly result from the strong disturbance caused by the gas evolution in droplets.

Figures 4–8 display the in-situ optical photos of tin electrodes during ECM tests under a droplet of 15 µL containing different concentrations of Na_2HPO_4 at various DC bias voltages. As shown in Figures 4a and 5a, prior to ECM, droplets of 15 µL on the surfaces of electrodes are in different shapes. The contact area of the droplet on the electrode surface cannot be accurately controlled in the WD test [31]. As shown in Figures 4–6, tree-like dendrites and white precipitates can be observed during ECM in WD tests without and with low concentrations of Na_2HPO_4 (such as 50 mg/L and 100 mg/L). Within the Na_2HPO_4 concentration range from 500 to 1000 mg/L, the dendrite cannot be observed while some white and dark products can be found on the surface of the anode. As shown in Figure 7, a thick layer of white precipitates can be observed on the surface of the anode at a concentration of 500 mg/L. A thin dark product layer can be seen from Figure 8 at a concentration of 1000 mg/L. The number of bubbles formed on the surface of the cathode was augmented with the increase of bias voltage.

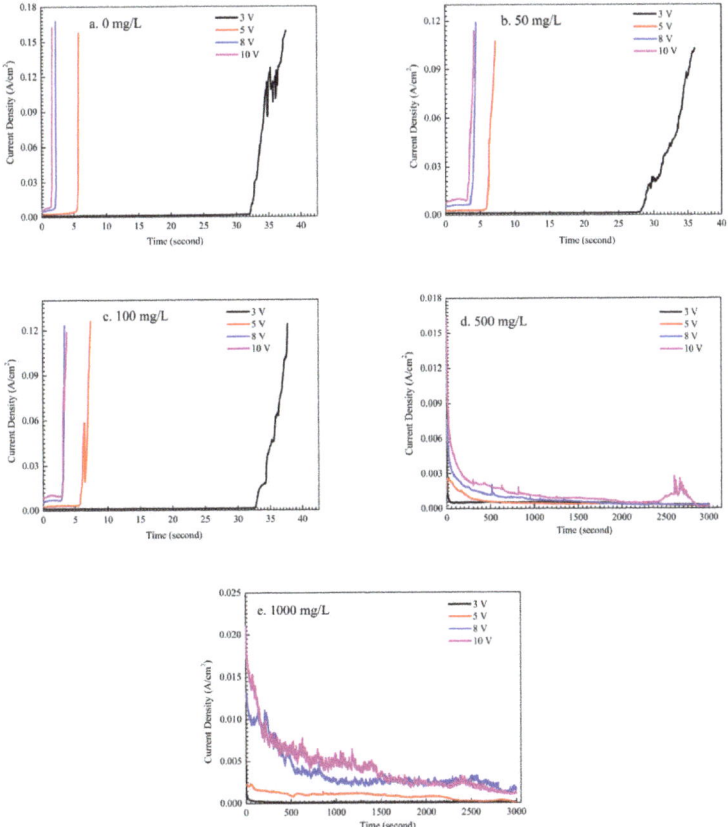

Figure 3. Current density vs. time curves for ECM tests of tin under a droplet of 15 µL containing different concentrations of Na_2HPO_4 under various bias conditions: (**a**) 0 mg/L; (**b**) 50 mg/L; (**c**) 100 mg/L; (**d**) 500 mg/L; (**e**) 1000 mg/L.

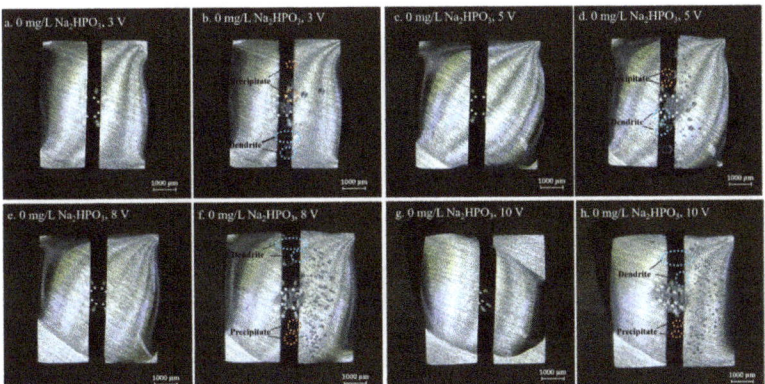

Figure 4. Optic photos for ECM tests of tin under a droplet of 15 µL containing 0 mg/L Na_2HPO_4 at various bias voltages (Anode is on the left and cathode is on the right).

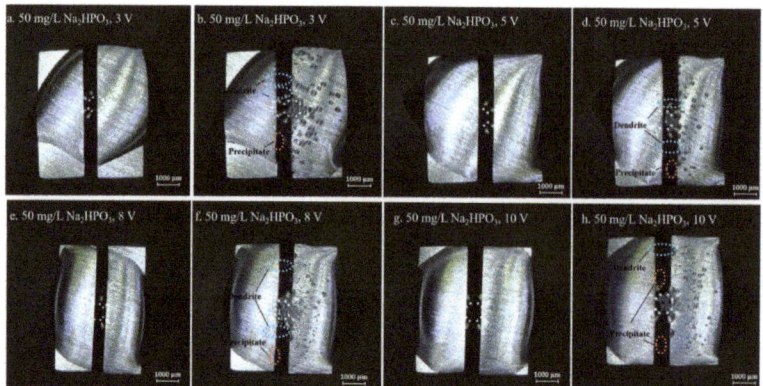

Figure 5. Optic photos for ECM tests of tin under a droplet of 15 μL containing 50 mg/L Na$_2$HPO$_4$ at various bias voltages (Anode is on the left and cathode is on the right).

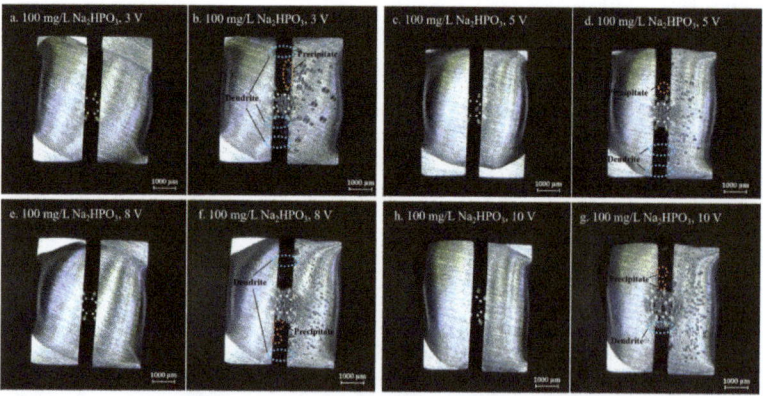

Figure 6. Optic photos for ECM tests of tin under a droplet of 15 μL containing 100 mg/L Na$_2$HPO$_4$ at various bias voltages (Anode is on the left and cathode is on the right).

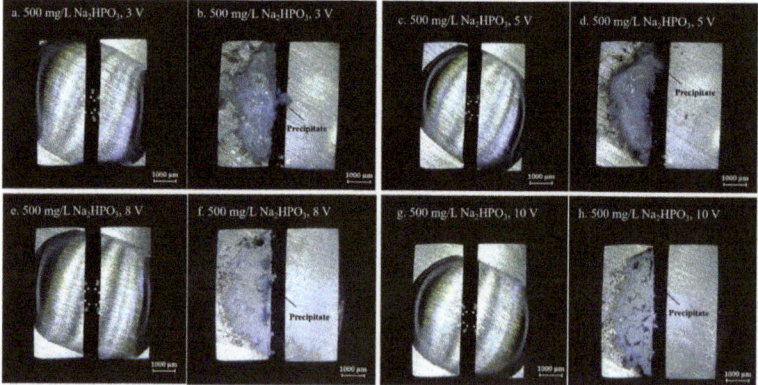

Figure 7. Optic photos for ECM tests of tin under a droplet of 15 μL containing 500 mg/L Na$_2$HPO$_4$ at various bias voltages (Anode is on the left and cathode is on the right).

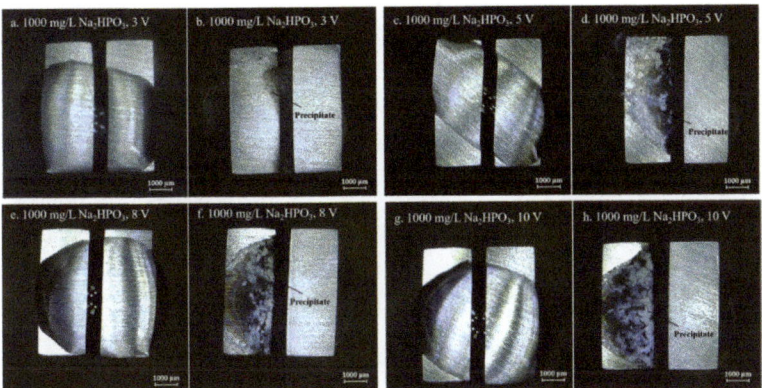

Figure 8. Optic photos for ECM tests of tin under a droplet of 15 µL containing 1000 mg/L Na$_2$HPO$_4$ at various bias voltages (Anode is on the left and cathode is on the right).

Figure 9 shows the microstructures of dendrites obtained after ECM tests under a droplet of 15 µL containing different concentrations of Na$_2$HPO$_4$ at 3 V. Fractal-like dendrites formed in the absence and presence of low concentrations of Na$_2$HPO$_4$ (for example, 50 mg/L and 100 mg/L) mainly consisting of tin element. Dendrites formed are all covered with precipitates. Upon addition of low concentrations of Na$_2$HPO$_4$, as shown in Figure 9c,e, branches of dendrites become more dense and tiny. Figure 10 presents the morphologies and corresponding EDS results of products formed on the anode after ECM tests under a droplet of 15 µL containing different concentrations of Na$_2$HPO$_4$ at 3 V. As shown in Figure 10a,b, some pits can be found on the surface of the anode, and the anodic products mainly consist of Sn (At% 34.72), O (At% 59.96) and Cl (At% 5.32) elements (Figure 10c). In the presence of 500 mg/L Na$_2$HPO$_4$, a thick layer of precipitate can be observed on the surface of the anode (Figure 10d,e), which is composed of Sn (At% 13.13), O (At% 78.90) and P (At% 7.97) elements (Figure 10f). Upon addition of 1000 mg/L Na$_2$HPO$_4$, a compact film can be observed from Figure 10g,h. The results of EDS (Figure 10i) indicate that this film consists of Sn (At% 20.54), O (At% 65.76) and P (At% 13.70) elements. The amount of P element of anodic products augments with the increasing Na$_2$HPO$_4$ concentration.

As is reported [32,33], the main anodic and cathodic reactions during ECM of tin in neutral NaCl solution are as follows:

Anodic reactions:

$$Sn \rightarrow Sn^{2+} + 2e^- \qquad (3)$$

$$Sn^{2+} \rightarrow Sn^{4+} + 2e^- \qquad (4)$$

Cathodic reactions:

$$2H_2O + 2e^- \rightarrow H_2 + 2OH^- \qquad (5)$$

$$Sn^{2+} + 2e^- \rightarrow Sn \qquad (6)$$

Hydration, adsorption and complex formation of tin ions [34] can occur in the anodic stage and HPO$_4^{2-}$ can determine the appearance of characteristic anodic regions [35]. In alkaline solution, the anodic products can consist of Sn^{2+}, Sn(OH)$^{3+}$, Sn(OH)$_2$, Sn(OH)$_4$, SnHPO$_4$, Sn(HPO$_4$)$_3^{4-}$, Sn$_3$(PO$_4$)$_2$, HSnO$_2^-$, SnO$_3^{2-}$, etc. (as shown in Reactions (8)–(12)) [36,37]. Polarization curves of tin in the absence and presence of Na$_2$HPO$_4$ are used to study the effect of Na$_2$HPO$_4$ on anodic behavior (Figure 11a). As shown in Figure 11b, the pitting potential (E_{pit}) of tin increases with the increasing concentration of Na$_2$HPO$_4$. For example, E_{pit} of tin increases from −194 mV (vs. SCE) to 1364 mV (vs. SCE) as concentration of Na$_2$HPO$_4$ increases from 0 mg/L to 1000 mg/L. E_{pit} is closely related to metal dissolution processes, and a higher E_{pit} means a lesser dissolution rate [38], indicating that the addition of Na_2HPO_4 can decrease the rate of anodic dissolution.

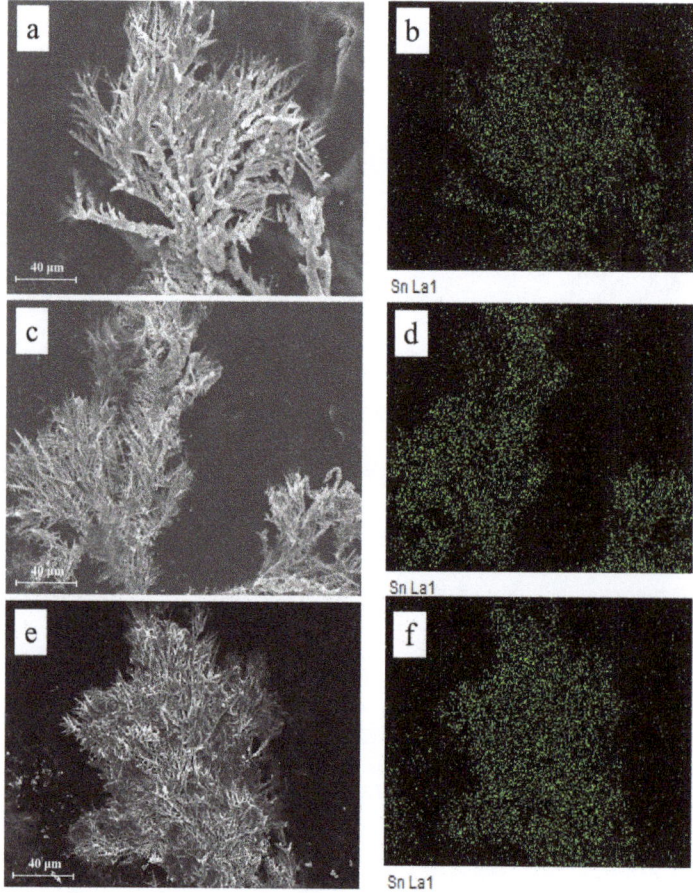

Figure 9. SEM images and corresponding EDS results of dendrites formed after ECM tests of tin under a droplet of 15 µL containing various concentrations of Na_2HPO_4 at 3 V: (**a**) 0 mg/L; (**b**) EDS-Mapping results of image (**a**); (**c**) 50 mg/L; (**d**) EDS-Mapping results of image (**c**); (**e**) 100 mg/L; (**f**) EDS-Mapping results of image (**e**).

As shown in Figure 10f,i, the anodic films formed after ECM mainly consist of Sn, O and P elements within the Na_2HPO_4 concentration range from 500 to 1000 mg/L. Compounds of tin phosphate can be formed during ECM. The following reactions are proposed to occur during the oxidation behavior of tin in Na_2HPO_4 solution [36]:

$$Sn + HPO_4^{2-} \rightleftharpoons SnHPO_4 + 2e^- \tag{7}$$

$$Sn^{2+} + HPO_4^{2-} \rightleftharpoons SnHPO_4 \text{ (aq)} \tag{8}$$

$$Sn^{2+} + HPO_4^{2-} \rightleftharpoons SnHPO_4 \text{ (s)} \tag{9}$$

$$Sn^{2+} + 3HPO_4^{2-} \rightleftharpoons Sn(HPO_4)_3^{4-} \tag{10}$$

$$3Sn^{2+} + 2PO_4^{3-} \rightleftharpoons Sn_3(PO_4)_2 \tag{11}$$

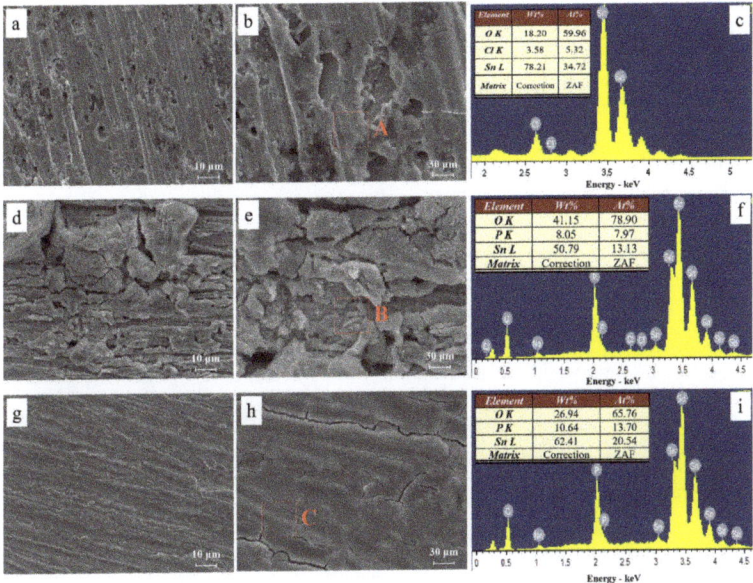

Figure 10. SEM images and corresponding EDS results of precipitates formed after ECM tests of tin under a droplet of 15 µL containing various concentrations of Na_2HPO_4 at 3 V: (**a**) 0 mg/L; (**b**) enlarged image of image (**a**); (**c**) EDS results of section A; (**d**) 500 mg/L; (**e**) enlarged image of image (**d**); (**f**) EDS results of section B; (**g**) 1000 mg/L; (**h**) enlarged image of image (**g**); (**i**) EDS results of section C.

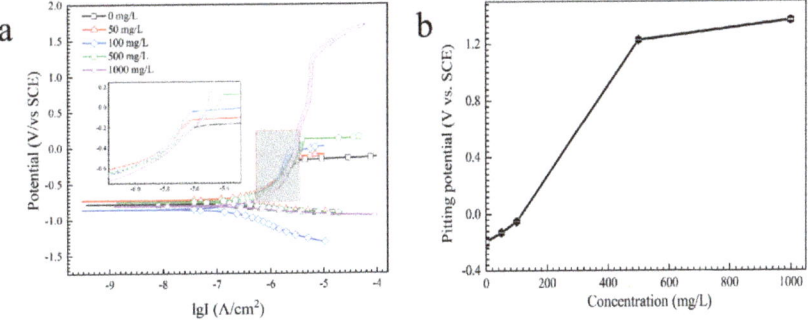

Figure 11. (**a**) Potentiodynamic polarization curves and (**b**) pitting potential of tin in the presence of various concentrations of Na_2HPO_4.

3.3. Effect of Applied DC Bias Voltage on the ECM of Tin

As shown in Figure 3, higher applied DC bias voltages indicated larger current densities during ECM tests. Mean TTF decreased with rising DC bias voltage, indicating that the increasing bias voltage accelerated the growth of dendrite. As the precondition for ECM, rates of anodic dissolution, metal migration and metal deposition (as shown in Reactions (3), (4) and (6)) largely depend on the DC bias voltage. As shown in Figures 4–8, the number of bubbles that formed on the surface of cathode increased at higher DC bias voltages, suggesting that the hydrogen evolution reaction (Reaction (5)) was also accelerated with the increasing bias voltage.

4. Conclusions

The inhibition effect of Na_2HPO_4 on the ECM of tin was evaluated by electrochemical methods and surface characterizations in this work. The following conclusions can be drawn:

(1) The growth rate of dendrite for tin increases with the increase of applied DC bias voltage. Dendrites formed after ECM tests in the absence and presence of low concentrations of Na_2HPO_4 (0–100 mg/L) have clearly a fractal-type structure and mainly consist of tin element.

(2) Within the relative higher Na_2HPO_4 concentration range (such as 500–1000 mg/L), owing to the formation of some insoluble compounds of tin phosphate on the surface of the anode, the rate of anodic dissolution is slowed down. The pitting potential of tin increases with the rising concentration of Na_2HPO_4 and the growth of the dendrite is retarded.

Author Contributions: B.L.: co-designed the project, characterization, data analysis and writing (review and editing); H.W.: co-designed the project, investigation and writing (original draft); S.W. and W.X.: provided data acquisition and analysis support; X.G.: co-designed the project and supervised the overall project. All authors have read and agreed to the published version of the manuscript.

Funding: This work was supported by the National Natural Science Foundation of China (Grant No. 51971067), Science and Technology Research Project of Guangzhou (No. 202002010007), Platform Research Capability Enhancement Project of Guangzhou University (Grant No. 69-620939), and Guangzhou University's 2020 Training Program for Talent (Grant No. 69-62091109).

Acknowledgments: The authors would like to acknowledge the financial from the National Natural Science Foundation of China (Grant No. 51971067), Science and Technology Research Project of Guangzhou (No. 202002010007), Platform Research Capability Enhancement Project of Guangzhou University (Grant No. 69-620939), and Guangzhou University's 2020 Training Program for Talent (Grant No. 69-62091109).

Conflicts of Interest: The authors declare no conflict of interest.

References

1. Kyeremateng, N.; Brousse, T.; Pech, D. Microsupercapacitors as miniaturized energy-storage components for on-chip electronics. *Nat. Nanotechnol.* **2017**. *12*, 7–15. [CrossRef]
2. Jin, S.K.; Kim, H.; Ryu, J.; Hahn, H.; Jang, S.; Joung, J.W. Inkjet printed electronics using copper nanoparticle ink. *J. Mater. Sci. Mater. Electron.* **2010**, *21*, 1213–1220.
3. Liao, B.; Cen, H.; Chen, Z.; Guo, X. Corrosion behavior of Sn-3.0Ag-0.5Cu alloy under chlorine-containing thin electrolyte layers. *Corros. Sci.* **2018**, *143*, 347–361. [CrossRef]
4. Qi, X.; Ma, H.; Wang, C.; Shang, S.; Li, X.; Wang, Y.; Ma, H. Electrochemical migration behavior of Sn-based lead-free solder. *J. Mater. Sci. Mater. Electron.* **2019**, *30*, 14695–14702. [CrossRef]
5. Guo, T.; Elshekh, H.; Yu, Z.; Yu, B.; Wang, D.; Kadhim, M.; Chen, Y.; Hou, W.; Sun, B. Effect of crystalline state on conductive filaments forming process in resistive switching memory devices. *Mater. Today Commun.* **2019**, *20*, 1–5. [CrossRef]
6. Niu, X.; Shen, L.; Chen, C.; Zhou, J.; Chen, L. An Arrhenius-type constitutive model to predict the deformation behavior of Sn-0.3Ag-0.7Cu under different temperature. *J. Mater. Sci. Mater. Electron.* **2019**, *30*, 14611–14620. [CrossRef]
7. Zou, S.; Li, X.; Dong, C.; Ding, K.; Xiao, K. Electrochemical migration, whisker formation, and corrosion behavior of printed circuit board under wet H2S environment. *Electrochim. Acta.* **2013**, *114*, 363–371. [CrossRef]
8. Gain, A.; Zhang, L. Effect of Ag nanoparticles on microstructure, damping property and hardness of low melting point eutectic tin-bismuth solder. *J. Mater. Sci. Mater. Electron.* **2017**, *28*, 15718–15730. [CrossRef]
9. Abtew, M.; Selvaduray, G. Lead-free solders in microelectronics. *Mater. Sci. Eng. R Rep.* **2000**, *27*, 95–141. [CrossRef]
10. Veleva, L.; Dzib-Pérez, L.; González-Sánchez, J.; Pérez, T.R. Initial stages of indoor atmospheric corrosion of electronics contact metals in humid tropical climate: Tin and nickel. *Rev. Metal. Madrid.* **2007**, *43*, 101–110. [CrossRef]
11. Minzari, D.; Jellesen, M.; Moller, P.; Ambat, R. On the electrochemical migration mechanism of tin in electronics. *Corros. Sci.* **2011**, *53*, 3366–3379. [CrossRef]

12. Jiang, S.; Liao, B.; Chen, Z.; Guo, X. Investigation of electrochemical migration of tin and tin-based lead-free solder alloys under chloride-containing thin electrolyte layers. *Int. J. Electrochem. Sci.* **2018**, *13*, 9942–9949. [CrossRef]
13. Zhong, X.; Qiu, Y.; Chen, Z.; Zhang, G.; Guo, X. In situ study the electrochemical migration of tin under unipolar square wave electric field. *J. Eectrochem. Soc.* **2013**, *160*, D495–D500. [CrossRef]
14. Zhong, X.; Zhang, G.; Guo, X. The effect of electrolyte layer thickness on electrochemical migration of tin. *Corros. Sci.* **2015**, *96*, 1–5. [CrossRef]
15. Yi, P.; Xiao, K.; Dong, C.; Zou, S.; Li, X. Effects of mould on electrochemical migration behaviour of immersion silver finished printed circuit board. *Bioelectrochemistry* **2017**, *119*, 203–210. [CrossRef]
16. Liao, B.; Jia, W.; Sun, R.; Chen, Z.; Guo, X. Electrochemical migration behavior of Sn-3.0Ag-0.5Cu solder alloy under thin electrolyte layers. *Surf. Rev. Lett.* **2019**, *26*, 185–208. [CrossRef]
17. Liao, B.; Wang, H.; Xiao, E.; Cai, Y.; Guo, X. Recent advances in method of suppressing dendrite formation of tin-based solder alloys. *J. Mater. Sci. Mater. Electron.* **2020**. Accepted. [CrossRef]
18. Abd El Haleem, S.M.; Abd El Wanees, S.; Abd El Aal, E.; Diab, A. Environmental factors affecting the corrosion behavior of reinforcing steel II. Role of some anions in the initiation and inhibition of pitting corrosion of steel in $Ca(OH)_2$ solutions. *Corros. Sci.* **2010**, *52*, 292–302. [CrossRef]
19. Bastos, A.; Ferreira, M.; Simões, A. Corrosion inhibition by chromate and phosphate extracts for iron substrates studied by EIS and SVET. *Corros. Sci.* **2006**, *48*, 1500–1512. [CrossRef]
20. van Soestbergen, M.; Erich, S.; Huinink, H.; Adan, O. Dissolution properties of cerium dibutylphosphate corrosion inhibitors. *Corros. Eng. Sci. Technol.* **2013**, *48*, 234–240. [CrossRef]
21. Alibakhshi, E.; Ghasemi, E.; Mahdavian, M. Corrosion inhibition by lithium zinc phosphate pigment, Corros. *Sci.* **2013**, *77*, 222–229.
22. Etteyeb, N.; Dhouibi, L.; Takenouti, H.; Alonso, M.; Triki, E. Corrosion inhibition of carbon steel in alkaline chloride media by Na_3PO_4. *Electrochim. Acta.* **2007**, *52*, 7506–7512. [CrossRef]
23. Awad, S.; Kassab, A. Behaviour of tin as metal-metal phosphate electrode and mechanism of promotion and inhibition of its corrosion by phosphate ions. *J. Electroanal. Chem.* **1969**, *20*, 203–212. [CrossRef]
24. Yohai, L.; Vázquez, M.; Valcarce, B. Phosphate ions as corrosion inhibitors for reinforcement steel in chloride-rich environments. *Electrochim. Acta.* **2013**, *102*, 88–96. [CrossRef]
25. Almobarak, N. Cyclic voltammetry study of passivation of tin in sodium dihydrogen phosphate solution. *Chem. Technol. Fuels Oils* **2012**, *48*, 321–330. [CrossRef]
26. Fulmer, M.; Brown, P. Hydrolysis of dicalcium phosphate dihydrate to hydroxyapatite. *J. Mater. Sci. Mater. Electron.* **1998**, *9*, 197–202. [CrossRef]
27. Zhong, X.; Zhang, G.; Qiu, Y.; Chen, Z.; Guo, X. Electrochemical migration of tin in thin electrolyte layer containing chloride ions. *Corros. Sci.* **2013**, *74*, 71–82. [CrossRef]
28. Liao, B.; Wei, L.; Chen, Z.; Guo, X. Na_2S-influenced electrochemical migration of tin in a thin electrolyte layer containing chloride ions. *RSC Adv.* **2017**, *7*, 15060–15070. [CrossRef]
29. Medgyes, B.; Rigler, D.; Illés, B.; Harsányi, G.; Gál, L. Investigating of electrochemical migration on low-Ag lead-free solder alloys. In Proceedings of the 18th International Symposium for Design and Technology of Electronic Packaging, Alba Iulia, Romania, 25–28 October 2012; pp. 147–150.
30. Liao, B.; Li, Z.; Cai, Y.; Guo, X. Electrochemical migration behavior of Sn-3.0Ag-0.5Cu solder alloy under SO_2 polluted thin electrolyte layers. *J. Mater. Sci. Mater. Electron.* **2019**, *30*, 5652–5661. [CrossRef]
31. Zhong, X.; Yu, S.; Chen, L.; Hu, J.; Zhang, Z. Test methods for electrochemical migration: A review. *J. Mater. Sci. Mater. Electron.* **2016**, *28*, 1–11. [CrossRef]
32. Liao, B.; Cen, H.; Chen, Z.; Guo, X. Effect of Organic Acids on the Electrochemical Migration of Tin in Thin Electrolyte Layer. *Innov. Corros. Mater. Sci. (Former. Recent Pat. Corros. Sci.)* **2019**, *9*, 74–84. [CrossRef]
33. Liao, B.; Chen, Z.; Qiu, Y.; Zhang, G.; Guo, X. Effect of citrate ions on the electrochemical migration of tin in thin electrolyte layer containing chloride ions. *Corros. Sci.* **2016**, *112*, 393–401. [CrossRef]
34. Cilley, W.A. Solubility of tin (II) orthophosphate and the phosphate complexes of tin (II). *Inorg. Chem.* **1968**, *7*, 612–614. [CrossRef]
35. Kamel, K.; Awad, S.; Kassab, A. Non-corrosive action of the tertiary phosphate ion on aluminium. *J. Electroanal. Chem.* **1981**, *127*, 195–202. [CrossRef]
36. Ciavatta, L.; Iuliano, M. Formation equilibria of tin (II) orthophosphate complexes. *Polyhedron* **2000**, *19*, 2403–2407. [CrossRef]

37. Duc, H.D.; Tissot, P. Anodic behaviour of tin in neutral phosphate solution. *Corros. Sci.* **1979**, *19*, 179–190. [CrossRef]
38. Britton, S. The Corrosion and Oxidation of Metals. *J. Electrochem. Soc.* **1977**, *12*, 5. [CrossRef]

© 2020 by the authors. Licensee MDPI, Basel, Switzerland. This article is an open access article distributed under the terms and conditions of the Creative Commons Attribution (CC BY) license (http://creativecommons.org/licenses/by/4.0/).

Article

Low Temperature Sealing Process and Properties of Kovar Alloy to DM305 Electronic Glass

Zhenjiang Wang [1], Zeng Gao [1,*], Junlong Chu [1], Dechao Qiu [1] and Jitai Niu [1,2,3]

1 School of Materials Science and Engineering, Henan Polytechnic University, Jiaozuo 454003, China; wangzhenjiang2017@163.com (Z.W.); chujunlong123@163.com (J.C.); qiu_dechao@163.com (D.Q.); niujitai@163.com (J.N.)
2 State Key Laboratory of Advanced Welding and Joining, Harbin Institute of Technology, Harbin 150001, China
3 Henan Jingtai High-Novel Materials Ltd. of Science and Technology, Jiaozuo 454003, China
* Correspondence: mrgaozeng@163.com or gaozeng@hpu.edu.cn; Tel.: +86-138-3913-4383

Received: 12 June 2020; Accepted: 10 July 2020; Published: 13 July 2020

Abstract: The low temperature sealing of Kovar alloy to DM305 electronic glass was realized by using lead-free glass solder of the Bi_2O_3-ZnO-B_2O_3 system in atmospheric environment. The sealing process was optimized by pre-oxidation of Kovar alloy and low temperature founding of flake glass solder. The effects of sealing temperature and holding time on the properties of sealing joint were studied by means of X-ray diffraction (XRD), scanning electron microscope (SEM), energy dispersive X-ray spectroscopy (EDS), etc. The results showed that the pre-oxidized Kovar alloy and DM305 electronic glass were successfully sealed with flake glass solder at the sealing temperature of 500 °C for 20 min. Meanwhile, the joint interface had no pores, cracks, and other defects, the shear strength was 12.24 MPa, and the leakage rate of air tightness was 8×10^{-9} Pa·m^3/s. During the sealing process, element Bi in glass solder diffused into the oxide layer of Kovar alloy and DM305 electronic glass about 1 µm, respectively.

Keywords: Kovar alloy; DM305 electronic glass; glass solder; low temperature sealing

1. Introduction

Glass materials have good light transmittance, insulation, and corrosion resistance, and many new applications and manufacturing processes will involve glass in combination with other materials [1]. Metal materials have good electrical conductivity, thermal conductivity, and plastic toughness. The comprehensive properties of glass and metal can be utilized through the combination of glass and metal. Glass to metal joints have been used in many applications. For example, glass insulation is required between the electronic package housing and the lead wire pin for microelectronic metal packaging [2–4]. A glass brazing material needs to be employed to seal between glass tube and metal pipe to achieve a certain degree of vacuum for parabolic trough receivers [5–8], and glass to metal sealing also plays a key role and can prevent the leakage of fuel and air in planar solid oxide fuel cells (SOFCs) [9–11]. Kovar alloy is a Fe-Ni-Co alloy with Fe as the main matrix element, and its expansion coefficient is similar to that of silicon-boron hard glass; therefore, Kovar alloy and glass sealing are widely used in the field of electronic packaging related to in-candescent lamps, electron tubes, and housing for semi-conductors [12]. However, Kovar alloy has high density, which is increasingly difficult to meet the light-weight requirements for modern electronic packaging [13]. Therefore, a new type of electronic packaging material is required. Aluminum metal matrix composites possess excellent performance such as light-weight, good thermal conductivity, and adjustable linear coefficient of thermal expansion, making them the first choice of the new generation electronic packaging materials [3,14,15].

As the current joining technology of new packaging materials is under development, it is extremely difficult to completely replace the traditional Kovar alloy with aluminum metal matrix composites in a short period of time. However, the use of a combination of aluminum metal matrix composites and Kovar alloy can be designed to achieve weight reduction of the packaging shell, as shown in Figure 1. For example, the phased array radar T/R module was generally made of Kovar alloy. If a high-volume fraction silicon carbide reinforced aluminum matrix composite (65% SiC_p/Al-MMCs) with a similar low expansion coefficient is employed, the phased array radar can get a weight loss of about 2/3 [16]. For the T/R module sealing component of electronic packaging shell and glass insulation terminal, a part of 65% SiC_p/Al-MMCs is utilized to replace Kovar alloy, which can achieve a significant weight reduction of the packaging shell and at the same time is of great performance promotion. In addition, it is also a great economic value. In the past, the sealing of Kovar alloy with glass-insulated terminals (DM305 electronic glass) was generally carried out in three steps, that were, decarburization and degassing in humid hydrogen environment with high temperature, surface pre-oxidation treatment and sintering with glass. The sealing process of Kovar alloy to glass at high temperatures has been extensively studied by scholars. Luo et al. [17] studied the wetting and spreading behavior of borosilicate glass on the surface of pre-oxidized Kovar alloy, and proposed three wetting stages: incubation period, reaction period and equilibrium period. Kuo et al. [12] realized the matching sealing of ASF series glass to pre-oxidized Kovar alloy under the protection of inert gas, and the sealing strength of ASF110 glass to Kovar alloy was about 3.9 MPa. The pre-oxidation of Kovar was a key factor for sealing. During the pre-oxidation of Kovar alloy, FeO, Fe_3O_4, Fe_2O_3 and $(Fe, Co, Ni)_3O_4$ would appear in the oxide layer [18,19], but the main oxide was FeO. After pre-oxidation, metallic oxide layer will be generated on the surface of Kovar alloy. As it is well known, the chemical bond type of metallic oxide is ionic bond. Meanwhile, the chemical bond of DM305 electronic glass is a mixture of ionic and covalent bond. As a consequence, the low temperature glass which had a similar chemical bond could be employed in theory as an intermediate solder to join Kovar alloy and DM305 electronic glass, which was expected to realize a new packaging process of metal and glass.

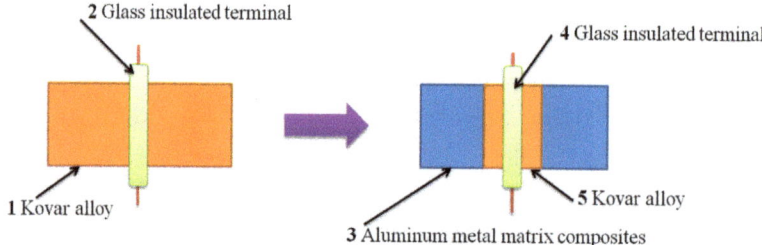

Figure 1. Schematic diagram of composite component formed by aluminum metal matrix composites replacing part of Kovar alloy.

The traditional sealing process is complicated and requires multiple high-temperature processes, wasting energy. In addition, high temperatures inevitably cause damage to the base metal of Kovar and glass-insulated terminals. More importantly, when the composite component involving aluminum metal matrix composites and Kovar alloy are sealed with glass-insulated terminal, it is no longer feasible to use traditional sintering at high temperature of 1000 °C since the melting point of aluminum is only 660 °C. New sealing methods are urgently needed. Due to environmental protection requirements, lead-free green glass powder will be used in large quantities replacing traditional lead-containing glass powder which is banned by many countries [20–24]. Simultaneously, some scholars have successfully combined sapphire, aluminum ceramics, Li–Ti ferrite, and other materials using lead-free glass solder in recent years [25–29]. In this study, low-temperature green glass powder Bi_2O_3-ZnO-B_2O_3 system which has similar chemical bonding with the pre-oxidized Kovar alloy surface and DM305 electronic glass was employed as a solder to achieve the sealing of Kovar alloy to DM305 electronic glass at low temperature, aiming to solve the problem of sealing aluminum metal matrix composites and Kovar alloy composite component with glass-insulated terminals, and to promote the application of aluminum composite materials in the field of electronic packaging. To use low-temperature green glass powder not only has a simple process and energy saving, but also responds to the world's call for green materials according to Waste Electrical and Electronic Equipment (WEEE) Directive and the Restriction of Hazardous Substances (RoHS) Directive in Electrical and Electronic Equipment [30].

2. Materials and Methods

2.1. Materials Preparation and Sealing Process

The specimens with the size of 15 × 10 × 2 mm were machined from the 4J29 Kovar alloy, whose composition was Ni 29.0 wt.%, Co 17.2 wt.%, Mn 0.3 wt.%, Si 0.2 wt.%, C 0.02 wt.%, with Fe balanced. Each surface of the flake samples was polished by 800 grade metallographic papers, and then ultrasonically cleaned in acetone for 10 min and in alcohol for 15 min, respectively. After being dried, the Kovar alloy samples were placed into a resistance furnace (GWL-1200), which was also employed for sealing test, and the temperature control accuracy was ±1 °C. The samples were firstly heated up to 900 °C at the rate of 15 °C/min and to be held for 1 min for pre-oxidation in air. The commercially available DM305 electronic glass had the same size as Kovar alloy. Its nominal composition was Al_2O_3 3.5 wt.%, B_2O_3 20.3 wt.%, K_2O 4.9 wt.%, Na_2O 3.8 wt.%, Fe_2O 30.1 wt.%, with SiO_2 balanced. The composition of Bi_2O_3-ZnO-B_2O_3 system glass solder utilized in the research is shown in Table 1. The coefficient of thermal expansion (CTE) of three categories of materials used in this work are shown in Table 2.

Table 1. Chemical compositions of glass solder (in wt.%).

Composition	Bi_2O_3	ZnO	B_2O_3	Na_2CO_3	SiO_2
Content	70–80	0–10	0–10	0–5	0–5

Table 2. Coefficient of thermal expansion of 4J29 Kovar alloy, DM305 electronic glass and glass solder.

Material	4J29 Kovar Alloy	DM305 Electronic Glass	Glass Solder
CET × 10^{-7}/°C	5.1–5.5	4.8–5.0	6.5–7.5

In order to avoid air pore defects caused by air gaps in the powder solder during the sealing process, the powder solder was putted in a self-designed mold first, as shown in Figure 2a. Then the mold was heated from room temperature to 200 °C at the rate of 10 °C/min and held for 10 min. Subsequently, it was heated to 460 °C at the rate of 10 °C/min and held for 30 min in the resistance furnace surrounded by atmosphere environment. After that, the glass column (φ5 × 3 mm) prepared for wetting experiment and flake glass solder (7.5 × 7.5 × 0.5 mm) used for sealing were obtained. The wetting experiment was carried out by classical sessile drop method. The glass column was placed in the middle of pre-oxidized Kovar and DM305 electronic glass primarily, then the specimens were heated directly up to 480, 500, 520, 540 and 560 °C at the rate of 10 °C/min and held for 30 min in the resistance furnace, respectively. The sandwich specimen assembled in a sealing fixture is shown in Figure 2b, the overlap length was 10 mm. During sealing process, the pressure of 0.01 MPa was applied to the sandwich specimen. The self-designed device in Figure 2c was used for joint shear test. The sealing experiments were performed by heating the specimens in the resistance furnace in air. Figure 3 shows the process of sealing. The sandwich specimen was firstly heated up to 250 °C at the rate of 10 °C/min and held for 10 min to keep temperature uniform and reduce thermal stress. Then the specimen was continually heated up to sealing temperature at the rate of 5 °C/min and held for 10~40 min. After that, the specimen was cooled down to 350 °C at the rate of 5 °C/min, and then was cooled down to room temperature in furnace. Sealing processes of Kovar alloy to DM305 electronic glass are listed in Table 3.

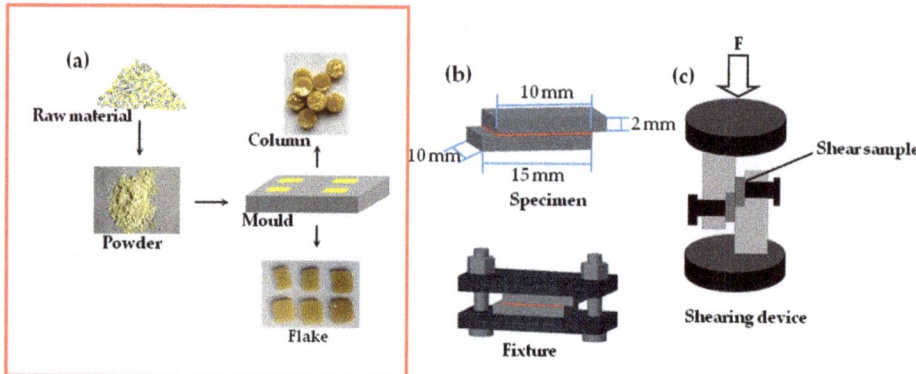

Figure 2. Schematic diagrams of experimental procedure: (**a**) Glass powder molding process, (**b**) sealing assembly, and (**c**) shear testing.

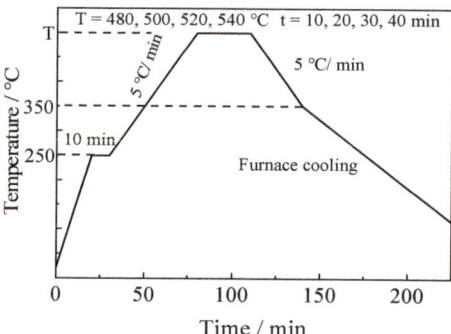

Figure 3. Process of sealing with different sealing temperatures and holding times.

Table 3. Sealing processes of Kovar alloy to DM305 electronic glass.

Sealing Temperature/°C	Holding Time/min	Sealing Pressure/MPa	Glass Solder
480	20	0.01	flake
500	20	0.01	flake
520	20	0.01	flake
540	20	0.01	flake
500	10	0.01	flake
500	30	0.01	flake
500	40	0.01	flake

2.2. Characterization Methods

Before the sealing experiments, thermophysical properties of the glass solder were analyzed by differential scanning calorimetry (DSC, Q100, TA Instruments, New Castle, DE, USA). Shearing test of the sealing joint was performed at a constant rate of 0.02 mm/min by using an electronic universal testing machine (CMT5105, MTS Systems (China) Co., Ltd., Shenzhen, China) at room temperature. Three samples were employed for each experiment condition, and the adopted shear strength was the average of the three samples. The wettability of the glass solder on base material was observed by metallographic microscope (OLYMPUS GX51, Olympus Corporation, Tokyo, Japan). The microstructure and elements composition of the joints were carried out by scanning electron microscope (SEM, Carl Zeiss NTS GmbH, Merlin Compact, Jena, Germany) coupled with energy dispersive X-ray spectroscopy (EDS). The phases in joint were analyzed by X-ray diffraction (XRD, Dmax- RB, Rigaku Corporation, Tokyo, Japan). To evaluate the effectiveness of sealing process, air tightness of the joint was measured with a ZQJ-530 helium leak mass spectrometer (KYKY Technology Development Ltd., Beijing, China). The final air tightness value was selected from the worst one among three tested samples for each experiment condition.

3. Results and Discussion

3.1. Thermophysical Properties of Bi_2O_3-ZnO-B_2O_3 Syetem Glass

The DSC curve of the glass solder is shown in Figure 4. The temperature and meaning of each feature point on the curve are shown in Table 4. According to the analysis of DSC curve, the glass transition temperature (T_g) of the glass solder is 350 °C and the softening temperature (T_f) is 429 °C, respectively. Based on Ma Yingren's research on low-melting glass, to fully wet the sealing interface of the base material, the viscosity of low-temperature glass must reach 10^3~10^5 Pa·s. To reach this viscosity, the sealing temperature was generally set to be higher than the softening point in a range of 50~100 °C [31]. As a consequence, the sealing temperature in this research was set to be in the range of 480~560 °C based on DSC analysis of Bi_2O_3-ZnO-B_2O_3 system.

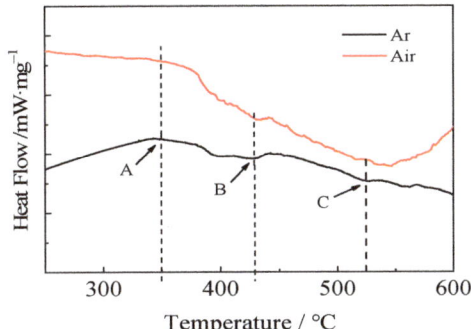

Figure 4. Differential scanning calorimetry (DSC) curve of the Bi_2O_3-ZnO-B_2O_3 system glass solder.

Table 4. Temperatures and meanings of the points on DSC curve in Figure 4.

Point	Temperature/°C	Meaning
A	350	Glass transition point
B	429	Glass softening point
C	520	Glass flow point

Within the heating process, the first obvious exothermic peak appeared when the temperature had reached about 440 °C, which indicated that crystallization may occur in the heating process. In order to further verify the above analysis, powdered and flaky glass solders were prepared for XRD analysis. Figure 5 shows that the powdered glass solder was amorphous, but a small amount of crystallization occurred in the flaky glass solder which was founded at 460 °C for 30 min, and the crystalline phase was mainly Bi_2O_3. As presented in Table 1, the content of Bi_2O_3 in glass solder was high. As a result, the amorphous phase of Bi_2O_3 in a thermodynamic metastable state partially crystallized more likely during the heating process.

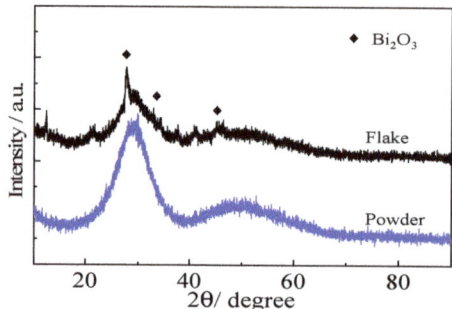

Figure 5. XRD patterns of the powder and flake glass solder.

The wetting angle reflects the bonding ability of the glass solder to base material. A reasonable sealing temperature range could be chosen according to the wetting angle of glass solder on two base materials at different temperatures. Figure 6 is a typical wetting physical and metallographic picture of the glass solder on both base materials at a temperature of 500 °C for 20 min. It can be seen that the glass solder had formed a good interface bonding with Kovar and DM305 electronic glass, respectively. The wetting angle on Kovar alloy was 38° which was smaller than that on DM305 electronic glass 65°, and both were less than 90°, indicating that this glass solder could wet the two of base materials and the sealing process was possible.

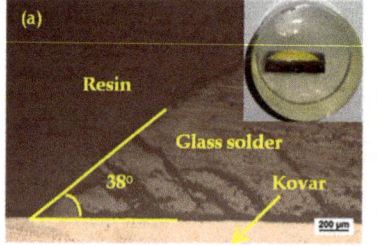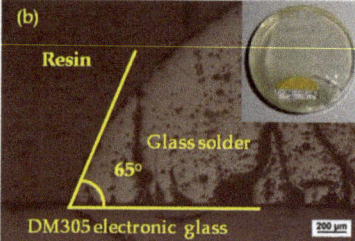

Figure 6. Typical wetting images of glass solder on base materials at 500 °C for 20 min: (a) Glass solder/Kovar and (b) glass solder/DM305 electronic glass.

Figure 7 shows the wetting angles of the glass solder on base materials at different temperatures for 20 min. It could be seen that the wetting angle of the glass solder on each material was larger than 90° at 480 °C, suggesting the glass solder had a poor wettability on each base material. When the temperature increased to 500 and 520 °C, the wetting angle decreased significantly as the viscosity of glass solder decreased rapidly as the temperature rose. During the temperature range of 540 to 560 °C, the wetting angle on electronic glass did not change significantly, but there was an obvious "collapse" for Kovar alloy at 560 °C whether it was pre-oxidized or not, and the wetting angle was reduced to 20°, reaching a minimum. Whether the Kovar alloy underwent pre-oxidation treatment had little effect on wetting angle when the temperature exceeded 540 °C. It could be explained that when the temperature continued to increase after 540 °C, the viscosity of glass solder decreased accordingly, and the change of viscosity was the main factor that promoted wetting, while the surface state of Kovar alloy had little effect on wetting. It was generally considered that the wetting angle was suitable for sealing in the range of 45° to 90° [32]. Therefore, it was speculated that the reasonable sealing temperature range was from 500 to 540 °C in this research.

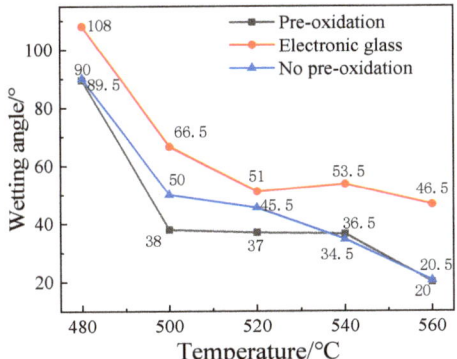

Figure 7. Wetting angle between glass solder and different base metals at different temperatures for 20 min.

3.2. Effect of Sealing Temperature on Sealing Joint Microstructure Evolution

In order to investigate the effect of sealing temperature on microstructure evolution of the sealing joints, the joints were sealed using the temperatures of 480, 500, 520 and 540 °C with a constant holding time of 20 min. The SEM micrographs of the joints are shown in Figure 8. Figure 8a shows the secondary electronic image of the sealing joint obtained under the condition of sealing temperature 480 °C. It could be seen that the bonding between DM305 electronic glass and Kovar alloy was achieved. There were a small number of pores in the joint. The primary reason was that the low sealing temperature could result in the difficulty of venting gas due to the poor fluidity of glass solder. When the sealing temperature was 500 °C, as can be seen in Figure 8b. The joint could be divided into three regions: zone 1 at the bonding area between DM305 electronic glass and glass solder, zone 2 in the middle of the sealing joint and zone 3 at the bonding area between Kovar alloy and glass solder. Additionally, there were three kinds of interface between base metal and glass solder: Kovar–oxide layer, oxide layer–glass solder and DM305 electronic glass–glass solder, and each interface was well bonded. In zone 1, the interface between DM305 electronic glass–glass solder was smooth, which was due to original small surface roughness of DM305 electronic glass which could not be melt during the sealing process for its high softening point. There were no obvious cracks and pores at the interface, and some fine gray-white crystals generated. Zone 2 was an obvious Bi_2O_3-ZnO-B_2O_3 glass solder, with a few small pores and a small amount of gray-white crystal phases. In zone 3, a black oxide film which bonded the glass solder to Kovar alloy together formed after pre-oxidation, and the interface of glass solder–oxide film combined well. There were fine gray crystals at the interface, without obvious cracks, pores, and other defects. The interface between oxide film–Kovar alloy was also compact, and there was a small amount of fine oxide embedded into the fringe area of Kovar alloy.

Figure 8c shows the sealing joint obtained at the sealing temperature of 520 °C. It could be found that there were obvious cracks on the Kovar–oxide layer interface, and the sizes of the pores in the center of the sealing joint were obviously increased. Meanwhile, the sizes of crystal phases were larger with an average diameter about 20 μm compared to the previous with an average diameter about 10 μm in Figure 8b. The number of crystal phases at interface on both sides decreased, while the number of crystalline phases in the center of the joint increased, which may be due to the partial melting of the crystalline phase at the interface. When the sealing temperature was 540 °C, as shown in Figure 8d, there were large pores that penetrated the entire joint due to high temperature which caused the glass solder to be over-burned. Accordingly, the pores decreased the performance of the sealing joint significantly.

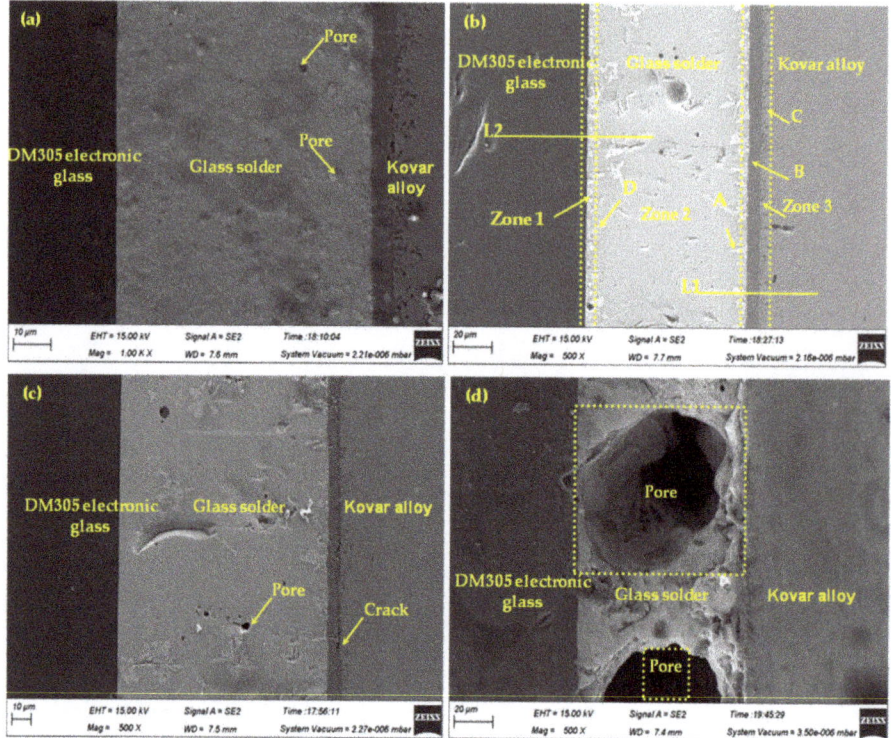

Figure 8. SEM images of the typical sealing joints at different sealing temperatures for 20 min: (**a**) 480 °C, (**b**) 500 °C, (**c**) 520 °C, and (**d**) 540 °C.

The point energy spectrum positions are shown in Figure 8b and the results and possible phases or matters in the joint are listed in Table 5. XRD analysis was performed to further confirm the phases in the joint. As can be seen, compared with the original glass powder shown in Figure 5, Bragg diffraction peaks could be also observed from the XRD pattern besides the amorphous peak, as shown in Figure 9. Therefore, the gray-white crystal phase A was $Bi_{24}B_2O_{39}$ according to the results of XRD analysis and the point energy spectrum, and this is consistent with the result in the previous literature [33]. It could be speculated that point B was mainly a FeO oxide layer of Kovar alloy, point C contained point FeO oxide and a matrix of Kovar alloy, and point D was glass solder, respectively, in Table 5.

Table 5. Energy dispersive X-ray spectroscopy (EDS) analysis of points at the interface of the sealing joint in Figure 8b (in wt.%).

Point	Fe	Co	Ni	O	Bi	B	Zn	Possible Phases or Matters
A	-	-	-	9.0	85.9	5.1	-	$Bi_{24}B_2O_{39}$
B	59.6	7.8	-	32.6	-	-	-	FeO
C	58.6	6.0	7.4	28.0	-	-	-	FeO, Kovar
D	-	-	-	10.1	82.8	4.1	3.0	Glass solder

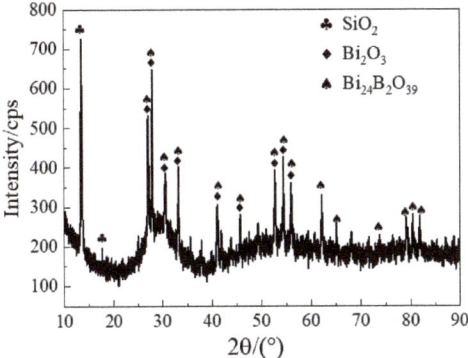

Figure 9. XRD patterns of the sealing joint in Figure 8b.

Simultaneously, line scanning analysis of the main elements was performed at marked positions in Figure 8b. At the L1 line scanning position, as shown in Figure 10a, the diffusion and dissolution layer were found to be relatively thin, about 1 µm. The thickness of oxide layer on Kovar alloy was about 6 µm. Due to the slight dissolution of oxide layer into glass solder, the element of Fe was distributed in the gradient of its diffusion and dissolution layer. At the same time, the element of Bi also diffused slightly from the glass solder layer into the oxide layer, about 1 µm. Similarly, at the DM305 electronic glass–glass solder interface, line scan analysis for the main elements was also performed at the marked position of line L2, as shown in Figure 10b. At the sealing temperature of 500 °C, only the glass solder melted, so the diffusion and dissolution layer were relatively thin. At the interface of DM305 electronic glass–the glass solder, there was a certain gradient distribution of Si, Bi, and O elements, and the diffusion distance of Si and O elements was about 2 µm, while that of Bi element was about 1 µm. It was noteworthy that there was a certain Bi element enrichment at the front of the interface of the glass solder–oxide layer and the same at the front of the interface of the glass solder–DM305 electronic glass, which was mainly due to the existence of Bi-rich crystalline phases at the front of the interface.

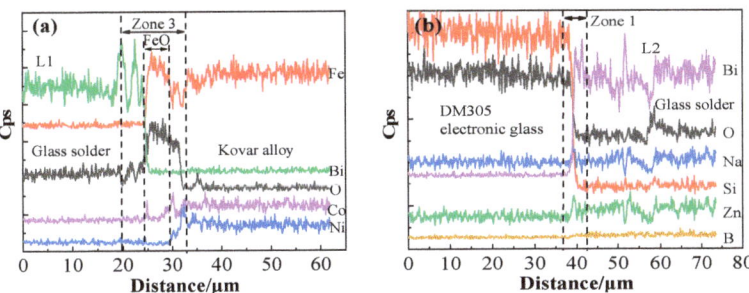

Figure 10. EDS line scanning of glass solder–Kovar alloy (a) and DM305 electronic glass–glass solder (b) of the sealing joint in Figure 8b.

3.3. Shear Tests of Sealing Joints

In order to find the best sealing process, the effects of sealing temperature and holding time on mechanical properties of sealing joints were studied, respectively, as shown in Figure 11a (fixed holding time 20 min) and Figure 11b (fixed sealing temperature 500 °C). As the sealing temperature increased, the joint shear strength increased firstly and then decreased, and reached a peak value of 12.24 MPa at 500 °C. This was mainly because the glass solder was not dense enough and not fully spread out at low temperature, since the viscosity of solder was too high and element diffusion ability at interface was weak. Meanwhile, there were a certain number of pores that were not expelled out in time in the joint, resulting in low interface bonding strength as well, as can be seen in Figure 8a. On the contrary, when the temperature was too high, the occurrence of glass solder over-burning caused enormous porosity, which reduced the shear strength too, as can be seen in Figure 8d.

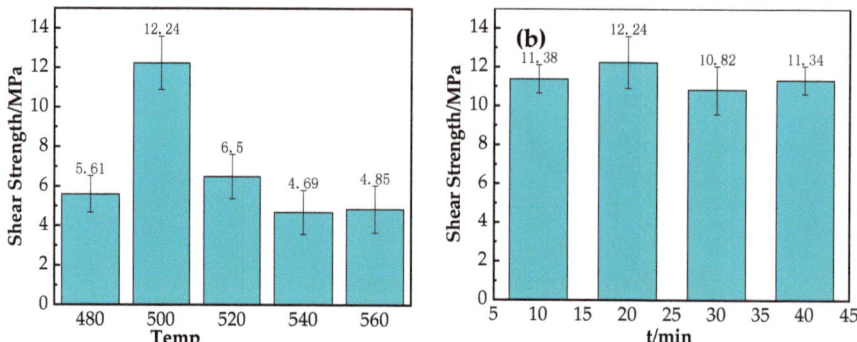

Figure 11. Effect of temperature (**a**) and holding time (**b**) on shear strength of the sealing joints.

Similarly, the changing trend of shear strength with the holding time changed was also studied, as shown in Figure 11b. In general, the shear strength did not change much with time; when the holding time was 20 min, the joint strength reached a peak value of 12.24 MPa. It could be speculated that the element diffusion was not sufficient when holding time was short. Instead, if the holding time was too long, the number and size of the crystalline phases would become more and larger, reducing the performance of the joint correspondingly. In order to further understand the mechanism of shear fracture, A typical force–displacement curve was displayed, as shown in Figure 12. The maximum shear load was 1256.29 N, and the shear strength was 12.56 MPa. The force–displacement curve was almost straight and no yield. Therefore, the sample belonged to brittle fracture with no obvious shear deformation.

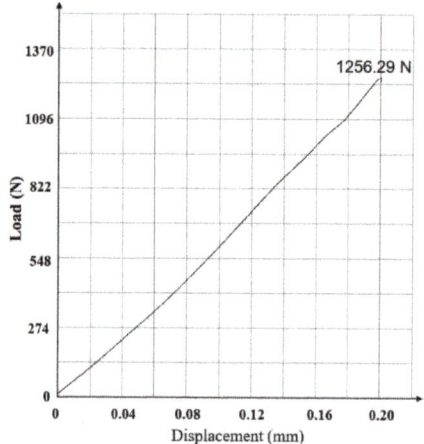

Figure 12. Typical force–displacement curve of the joint made at 500 °C for 20 min.

3.4. Air Tightness Tests of Sealing Joints

Air tightness is one of the most important factors affecting electronic packaging devices. By testing air tightness of the joint, the bonding condition of sealing interface can be evaluated from the air tightness, and it can also reflect the compactness of the oxide layer of Kovar alloy to a certain extent.

The air tightness of the sample is up to standard if the air leakage rate is less than 1×10^{-7} Pa·m^3/s according to Chinese national standard (GB 5594. 1-85, Air Tightness Test Method). Table 6 lists the air tightness of the sealing joint under different sealing temperatures for 20 min.

Table 6. Air tightness of samples at different temperatures for 20 min.

Temperatures/°C	480	500	520	540	560
Leakage rate/Pa·m^3/s	9×10^{-8}	8×10^{-9}	7×10^{-8}	6×10^{-7}	4×10^{-7}

The air tightness of the samples was up to standard when the sealing temperature was from 480 to 520 °C, but it was not qualified when the sealing temperature was higher than 520 °C, because the bonding temperature was too high and a large number of pores appeared in the solder. Meanwhile, the effect of holding time on the air tightness of the samples was also studied. It was found that the air tightness of all samples could meet the application requirements when the sealing temperature was 500 °C for 10~40 min, as shown in Table 7, indicating that the holding time had little effect on the air tightness of the samples.

Table 7. Air tightness of samples with different times at 500 °C.

Times/min	10	20	30	40
Leakage rate/Pa.m^3/s	5×10^{-8}	8×10^{-9}	6×10^{-8}	8×10^{-8}

4. Conclusions

In this work, Kovar alloy and DM305 electronic glass were successfully sealed by lead-free low temperature flake glass solder in atmospheric environment. Thermophysical properties, microstructure, shear testing, and air tightness testing of the joints were analyzed to confirm the performance of the sealing joints. This new low temperature sealing process can provide some technical reference and theoretical value for solving the problem of sealing aluminum metal matrix composites and Kovar alloy composite component to glass-insulated terminals, promoting the application of aluminum composite materials in the field of electronic packaging to some extent. The main results are as follows:

(1) The glass transition temperature T_g was 350 °C and softening temperature T_f was 429 °C for the Bi_2O_3-ZnO-B_2O_3 system lead-free glass solder. When the holding time was unchanged for 20 min, the wetting angle of the glass solder on both base materials was less than 90° after the temperature reached 500 °C, whereas the wetting angle was larger than 90° at 480 °C. Therefore, the glass solder had a good wettability on both base materials when the temperature was higher than 500 °C.

(2) The Bi_2O_3-ZnO-B_2O_3 system lead-free glass solder can realize the low temperature sealing of Kovar alloy to DM305 electronic glass. When the sealing temperature increased from 480 to 560 °C, the joint shear strength increased firstly and then decreased gradually. The maximum joint shear strength of 12.24 MPa can be reached when the sealing temperature of 500 °C and holding time of 20 min were utilized. Simultaneously, the joint air tightness could meet the application requirements.

(3) During the sealing process, crystallization of the Bi_2O_3-ZnO-B_2O_3 system glass solder appeared, and the crystal phases included $Bi_{24}B_2O_{39}$, Bi_2O_3, and SiO_2. At the sealing temperature of 500 °C, the crystallization mainly occurred at the interface between the glass solder and the base material. When the temperature reached to 520 °C, the crystallization phases at the interface decreased and gradually were transferred to the center of the joint.

(4) The mechanism of sealing mainly depends on short-range diffusion of some elements, such as Fe, Si, Bi and O elements, at the interface that lead to formation of an effective bonding between the glass solder and the base material according to line scan analysis.

Author Contributions: Investigation, Z.W., Z.G. and J.C.; resources, J.C. and D.Q.; supervision, J.N.; writing—original draft preparation, Z.W.; writing—review and editing, Z.G. and J.N. All authors have read and agreed to the published version of the manuscript.

Funding: The research was financially supported by the National Natural Science Foundation of China (No.51245008), the Science and Technology Project of Henan Province, China (No. 202102210036), the Fundamental Research Funds for the Universities of Henan Province, China (No. NSFRF180405).

Conflicts of Interest: The authors declare no conflict of interest.

References

1. Axinte, E. Glasses as engineering materials: A review. *Mater. Des.* **2011**, *32*, 1717–1732. [CrossRef]
2. Chanmuang, C.; Naksata, M.; Chairuangsri, T.; Jain, H.; Lyman, C. Microscopy and strength of borosilicate glass-to-Kovar alloy joints. *Mater. Sci. Eng. A* **2008**, *474*, 218–224. [CrossRef]
3. Wang, P.; Xu, D.; Niu, J. Vacuum brazing of electroless Ni–P alloy-coated SiCp/Al composites using aluminum-based filler metal foil. *Appl. Phys. A* **2016**, *122*, 122. [CrossRef]
4. Staff, M.; Fernie, J.A.; Mallinson, P.M.; Whiting, M.; Yeomans, J. Fabrication of a Glass-Ceramic-to-Metal Seal Between Ti-6Al-4V and a Strontium Boroaluminate Glass. *Int. J. Appl. Ceram. Technol.* **2016**, *13*, 956–965. [CrossRef]
5. Joshi, R.; Chhibber, R. Development and interface characterization of unmatched glass-metal joint. *J. Manuf. Process.* **2018**, *31*, 787–800. [CrossRef]
6. Joshi, R.; Chhibber, R. Failure Study of Compression Glass-Metal Joint for Parabolic Trough Receiver Tube application. *Mater. Today Proc.* **2018**, *5*, 14847–14851. [CrossRef]
7. Joshi, R.; Chhibber, R. High temperature wettability studies for development of unmatched glass-metal joints in solar receiver tube. *Renew. Energy* **2018**, *119*, 282–289. [CrossRef]
8. Lei, D.; Wang, Z.; Li, J.; Li, J.; Wang, Z. Experimental study of glass to metal seals for parabolic trough receivers. *Renew. Energy* **2012**, *48*, 85–91. [CrossRef]
9. Yan, J.; Wu, Y.; Lin, D.; Zhan, H.; Zhuang, H.; Sa, B.; Zhang, T. Structural transformation-induced surface strengthening of borosilicate sealing glass for solid oxide fuel cells. *Ceram. Int.* **2019**, *45*, 15629–15635. [CrossRef]
10. Smeacetto, F.; De Miranda, A.; Ventrella, A.; Salvo, M.; Ferraris, M. Shear strength tests of glass ceramic sealant for solid oxide fuel cells applications. *Adv. Appl. Ceram.* **2015**, *114*, 70. [CrossRef]
11. Javed, H.; Sabato, A.G.; Dlouhy, I.; Halasova, M.; Bernardo, E.; Salvo, M.; Herbrig, K.; Walter, C.; Smeacetto, F. Shear Performance at Room and High Temperatures of Glass–Ceramic Sealants for Solid Oxide Electrolysis Cell Technology. *Materials* **2019**, *12*, 298. [CrossRef] [PubMed]
12. Kuo, C.-H.; Cheng, P.-Y.; Chou, C.-P. Matched glass-to-Kovar seals in N_2 and Ar atmospheres. *Int. J. Miner. Met. Mater.* **2013**, *20*, 874–882. [CrossRef]
13. Zhang, S.; Xu, X.; Lin, T.; He, P. Recent advances in nano-materials for packaging of electronic devices. *J. Mater. Sci. Mater. Electron.* **2019**, *30*, 13855–13868. [CrossRef]
14. Wang, P.; Xu, D.; Zhai, Y.; Niu, J. The dissimilar brazing of Kovar alloy to SiCp/Al composites using silver-based filler metal foil. *Appl. Phys. A* **2017**, *123*, 569. [CrossRef]
15. Garg, P.; Jamwal, A.; Kumar, D.; Sadasivuni, K.K.; Hussain, C.M.; Gupta, P. Advance research progresses in aluminium matrix composites: Manufacturing & applications. *J. Mater. Res. Technol.* **2019**, *8*, 4924–4939. [CrossRef]
16. Wang, P.; Gao, Z.; Li, J.; Cheng, D.; Niu, J. Research on reaction brazing of Ti layer-coated SiCp/Al composites using Al-based filler metal foil. *Compos. Interfaces* **2019**, *26*, 1057–1068. [CrossRef]
17. Luo, D.; Shen, Z. Wetting and spreading behavior of borosilicate glass on Kovar. *J. Alloy. Compd.* **2009**, *477*, 407–413. [CrossRef]
18. Chern, T.-S.; Tsai, H.-L. Wetting and sealing of interface between 7056 Glass and Kovar alloy. *Mater. Chem. Phys.* **2007**, *104*, 472–478. [CrossRef]
19. Yates, P.M.; Mallinson, C.; Mallinson, P.M.; Whiting, M.; Yeomans, J. An Investigation into the Nature of the Oxide Layer Formed on Kovar (Fe–29Ni–17Co) Wires Following Oxidation in Air at 700 and 800 °C. *Oxid. Met.* **2017**, *88*, 733–747. [CrossRef]
20. Wang, F.; Dai, J.; Shi, L.; Huang, X.; Zhang, C.; Li, X.; Wang, L. Investigation of the melting characteristic, forming regularity and thermal behavior in lead-free V_2O_5–B_2O_3–TeO_2 low temperature sealing glass. *Mater. Lett.* **2012**, *67*, 196–198. [CrossRef]

21. Fredericci, C.; Yoshimura, H.; Molisani, A.; Fellegara, H. Effect of TiO2 addition on the chemical durability of Bi_2O_3–SiO_2–ZnO–B_2O_3 glass system. *J. Non-Crystalline Solids* **2008**, *354*, 4777–4785. [CrossRef]
22. He, F.; Wang, J.; Deng, D. Effect of Bi2O3 on structure and wetting studies of Bi_2O_3–ZnO–B_2O_3 glasses. *J. Alloy. Compd.* **2011**, *509*, 6332–6336. [CrossRef]
23. Hong, J.; Zhao, D.; Gao, J.; He, M.; Li, H.; He, G. Lead-free low-melting point sealing glass in SnO–CaO–P_2O_5 system. *J. Non-Crystalline Solids* **2010**, *356*, 1400–1403. [CrossRef]
24. Yang, Y.; Wu, L.; Teng, Y.; Jiang, F.; Lei, J.; Chen, H.; Cheng, J. Effect of ZnO content on microstructure, crystallization behavior, and thermal properties of xZnO-30B_2O_3-(65-x)Bi_2O_3-5BaO glass. *J. Non-Crystalline Solids* **2019**, *511*, 29–35. [CrossRef]
25. Guo, W.; Fu, L.; He, P.; Lin, T.; Wang, C. Crystallization and wetting behavior of bismuth–borate–zinc glass and its application in low temperature joining alumina ceramics. *J. Manuf. Process.* **2019**, *39*, 128–137. [CrossRef]
26. Guo, W.; Wang, T.; Lin, T.; Guo, S.; He, P. Bismuth borate zinc glass braze for bonding sapphire in air. *Mater. Charact.* **2018**, *137*, 67–76. [CrossRef]
27. Lin, P.; Lin, T.; He, P.; Guo, W.; Wang, J. Investigation of microstructure and mechanical property of Li–Ti ferrite/Bi_2O_3–B_2O_3–SiO_2 glass/Li–Ti ferrite joints reinforced by $FeBi_5Ti_3O_{15}$ whiskers. *J. Eur. Ceram. Soc.* **2015**, *35*, 2453–2459. [CrossRef]
28. Lin, P.; Lin, T.; He, P.; Sekulic, D.P. Wetting behavior and bonding characteristics of bismuth-based glass brazes used to join Li-Ti ferrite systems. *Ceram. Int.* **2017**, *43*, 13530–13540. [CrossRef]
29. Lin, P.; Wang, C.; He, P.; Wang, T.; Guo, W.; Lin, T. Brazing of Al_2O_3 ceramics by Bi_2O_3-B_2O_3-ZnO glass. *Procedia Manuf.* **2019**, *37*, 261–266. [CrossRef]
30. Zhang, L.; Han, J.-G.; He, C.-W.; Guo, Y.-H. Reliability behavior of lead-free solder joints in electronic components. *J. Mater. Sci. Mater. Electron.* **2012**, *24*, 172–190. [CrossRef]
31. Ma, Y. Sealing glass seven-low melting glass. *Glass Enamel.* **1993**, *21*, 50–54. (In Chinese)
32. Wang, J.; He, F.; Deng, D.; Tan, M. Research on Bismuth Glass and Stainless Steel Wettability. *J. Wuhan Univ. Technol.* **2010**, *32*, 60–64. [CrossRef]
33. He, F.; Wang, J.; Deng, D.; Cheng, J. Sintering Behavior of Bi_2O_3–ZnO–B_2O_3 System Low-melting sealing Glass. *J. Chin. Ceram. Soc* **2009**, *37*, 1791–1795. [CrossRef]

© 2020 by the authors. Licensee MDPI, Basel, Switzerland. This article is an open access article distributed under the terms and conditions of the Creative Commons Attribution (CC BY) license (http://creativecommons.org/licenses/by/4.0/).

Article

Solderability, Microstructure, and Thermal Characteristics of Sn-0.7Cu Alloy Processed by High-Energy Ball Milling

Ashutosh Sharma [1], Min Chul Oh [1,2], Myoung Jin Chae [1], Hyungtak Seo [1,*] and Byungmin Ahn [1,*]

1. Department of Energy Systems Research and Department of Materials Science and Engineering, Ajou University, Suwon 16499, Korea; ashu@ajou.ac.kr (A.S.); minlovehyo@ajou.ac.kr (M.C.O.); whe668@ajou.ac.kr (M.J.C.)
2. Metals Forming Technology R&D Group, Korea Institute of Industrial Technology, Incheon 21999, Korea
* Correspondence: hseo@ajou.ac.kr (H.S.); byungmin@ajou.ac.kr (B.A.); Tel.: +82-31-219-3532 (H.S.); +82-31-219-3531 (B.A.); Fax: +82-31-219-1613 (B.A.)

Received: 16 February 2020; Accepted: 10 March 2020; Published: 13 March 2020

Abstract: In this work, we have investigated the role of high-energy ball milling (HEBM) on the evolution of microstructure, thermal, and wetting properties of an Sn-0.7Cu alloy. We ball-milled the constituent Sn and Cu powders in eutectic composition for 45 h. The microstructural studies were carried out using optical and scanning electron microscopy. The melting behavior of the powder was examined using differential scanning calorimetry (DSC). We observed a considerable depression in the melting point of the Sn-0.7Cu alloy (≈7 °C) as compared to standard cast Sn-0.7Cu alloys. The resultant crystallite size and lattice strain of the ball-milled Sn-0.7Cu alloy were 76 nm and 1.87%, respectively. The solderability of the Sn-0.7Cu alloy was also improved with the milling time, due to the basic processes occurring during the HEBM.

Keywords: high-energy ball milling; soldering; joining; microstructure; interface

---

1. Introduction

Conventional Pb-bearing solders (Sn-37Pb, and Sn-40Pb) have been used extensively in electronic interconnects and microelectronic packaging devices for decades. The popularity of Pb-bearing solders is mainly due to its low cost and attractive properties required for soldering, e.g., low melting point, excellent wetting, and the absence of undesirable reaction compounds [1,2]. However, there are various regulations over the usage of Pb-bearing solder alloys due to their inherent toxicity to the human environment [3,4].

The last few decades have witnessed numerous research activities to find a substitute for Pb-bearing solders. Various Pb-free solders have been developed that are binary, ternary, or multicomponent alloys based primarily on Sn [5,6]. The most popular Pb-free solder is based on a ternary Sn-Ag-Cu (SAC) alloy [7]. SAC alloys have served in microelectronic packaging industries as a potential alternative to Sn-Pb alloys. However, there are still a few technology-related issues to consider when using SAC alloys [8]. The major concern in the use of Ag-bearing solder alloys is their higher melting point compared to Sn-Pb solders [9]. Another drawback is the high price of Ag across the globe. In addition, the reliability of Pb-free solders to various metallic substrates is still a big challenge. The presence of brittle Ag_3Sn intermetallic compounds (IMCs) in solders containing Ag deteriorates the joint properties [10]. Moreover, these popular SAC alloys, used in electronic packaging, contain additional Cu_6Sn_5, Cu_3Sn, etc., IMCs [8–10]. Other important Pb-free alloys include components like Cu, Bi, Zn, In, and Sb. Some of these have shown promising results, like lower melting points than Sn-Pb; however, the inherent brittleness of Bi, the scarcity of In in the earth's crust, the mild toxicity of Sb, and the oxidation issues associated with Zn have prevented their application [7,8,11–16]. Among these binary alloys, the Sn-Cu alloy seems to be a great candidate because of its cheaper price, absence of harmful Ag$_3$Sn IMCs, and good wetting. The eutectic Sn-0.7Cu alloy is widely used in wave soldering; however, it is not usually used in reflow soldering due to its poor strength and higher melting point than SAC alloys [17,18].

Therefore, melting point depression is of utmost importance in electronics packaging. Low-melting-point solder alloys may ease wetting to common metal substrates in electronic applications [19]. It is also established that grain size refinement allows for a possible depression in melting temperature. The grain boundary's excess energy acts as a driving force that causes a decrease in the enthalpy of melting. A disordered interface is believed to promote a depression in melting enthalpy. Thus, microstructural modification of solder alloys at nanoscale represents a good option for tailoring the melting point, in addition to mechanical strength [20–22]. Other methods include the use of alloying, or the addition of nanoparticles inside the soft metal matrix to enhance the melting point depression [23–25]. However, the use of secondary reinforcements has shown segregation and issues of poor wetting [24,25].

Common methods of alloy fabrication include sol-gel, chemical methods, vapor deposition techniques, melting and casting, severe plastic deformation, and high-energy ball milling (HEBM). Among these, HEBM has emerged as one of the exciting methods for the development of advanced nanostructured alloys [26]. HEBM is a potential method to produce low-melting-point nanocrystalline materials through continuous nanostructuring of metal powders. In addition, the high fraction of grain boundaries obtained during the milling process can also strengthen the mechanical and joint properties of materials when reflowed at higher temperatures [26–28].

There are various reports on the development of Pb-free solders via HEBM. Huang and his co-workers mechanically alloyed Sn-Zn and Sn-Sb alloy by HEBM [29]. Lai et al. produced Sn-Ag and Sn-Ag-Bi alloys by HEBM. They also obtained an enhancement of the wetting and soldering properties of these alloys [30]. Other researchers have fabricated Pb-free solders with several ceramic nano-reinforcements, e.g., ZrO$_2$, Al$_2$O$_3$, SiC, SnO$_2$, TiO$_2$, CNT, and graphene, through HEBM [24,25,31–35]. The present literature provides limited information on the lowering of the melting point of the existing Sn-0.7Cu alloy and the improvement of its mechanical strength.

Inspired by the aforementioned discussion, we have attempted to address the above-mentioned literature gaps. We fabricated an Sn-0.7Cu alloy by HEBM for 45 h. The melting properties of Sn-0.7Cu were evaluated after the milling. To assess the wetting properties of the Sn-0.7Cu alloy, the spreading behavior of the milled Sn-0.7Cu powders on copper substrates was also calculated.

2. Materials and Methods

The materials used for the HEBM were Sn and Cu powder (99.95% purity, Loba Chemie, Bangalore, India). The HEBM was performed at 300 rpm for 45 h in tungsten carbide (WC) vials containing 50 WC balls. The ball to powder weight ratio was 10:1. Stearic acid (0.2 wt. %) was mixed with the powder to avoid cold welding of the powder particles during the HEBM process. Table 1 shows the composition of the Sn-0.7Cu alloy milled for 45 h.

Table 1. Selected composition of the Sn-Cu alloy.

Element	Composition (wt. %)	
	Cu	Sn
Values	0.7	99.3

The Sn-0.7Cu powder was characterized by X-ray diffraction (XRD, Rigaku, Miniflex, Rigaku Corporation, Tokyo, Japan) to investigate the phase and structural evolution during the HEBM. The samples were prepared by mounting powder on glass holder slides with dimensions of 10 mm × 10 mm × 2 mm. The XRD was done at operating parameters of 35 kV and 35 mA with Cu K_α radiation (λ= 1.54056 Å). The samples were scanned from 25° to 70°. The crystallite size and lattice strain induced in the powder particles during HEBM were calculated using the Williamson-Hall (W-H) equation method [36].

The microstructural studies were done by scanning electron microscopy (SEM, Hitachi-4800, Tokyo, Japan). Prior to the SEM studies, the samples were handled and kept inside a glove box filled with Ar gas. The composition of the powder was identified using energy-dispersive X-ray spectroscopy (EDS, INCA, Oxford Instruments, Oxfordshire, UK).

To further study the alloying process, powder samples milled for various times were compacted into small pellets (Φ = 4mm) at 250 Pa [37]. The compacted samples were then mounted and tested for microhardness after metallographic polishing of up to 1 µm. The microhardness tests were performed by Vickers microhardness tester (Mitutoyo HM 200 series, Mitutoyo Corporation, Kanagawa, Japan) at 25 gf and 20 s. The microhardness measurements were repeated five times, and the average value was reported.

The melting point of the milled powders for various times was measured using a DSC (Diamond DSC, Perkin-Elmer, Waltham, MA, USA). The analysis was done from 50–350 °C at a heating rate of 10 °C/min. The samples were heated in an alumina pan under N_2 atmosphere. Assuming no weight gain or loss during heating, the enthalpy of melting (in J/g) was calculated by area under the curve divided by the sample's weight.

The wetting of the HEBM-ed Sn-0.7Cu powder samples was determined by the spread ratio method. The powder samples were reflowed at 250 °C on copper substrates. The change in the area spread ratio before and after melting was calculated.

3. Results and Discussion

The structural characterization of the phases evolved after milling the Sn-0.7Cu powder for 45 h, as shown in Figure 1.

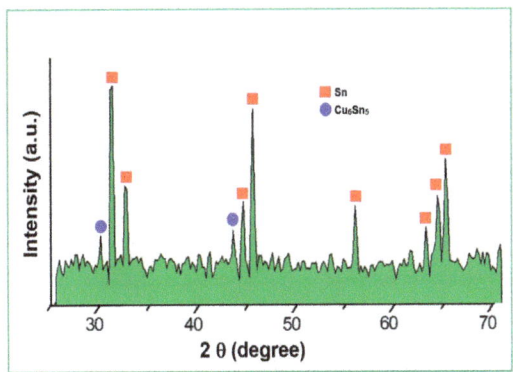

Figure 1. XRD pattern of the Sn-0.7 wt. % Cu powder after milling for 45 h.

The XRD diagram indicated the presence of tetragonal β-Sn. A few instances of Cu_6Sn_5 IMCs were also observed due to the diffusion of Cu atoms across the grain boundary. We also calculated the crystallite size and lattice strain of the Sn-0.7Cu alloy using the W-H equation (Table 2). According to the W-H equation [36], the peak broadening is related to the crystallite size (D), lattice strain (ε), full width at half maximum (β_{hkl}), and Bragg angle (θ) by

$$\beta_{hkl} \cos\theta = \frac{K\lambda}{D} + 4\varepsilon \sin\theta \tag{1}$$

where $K = 0.9$; the shape factor (assuming Cauchy—distribution), and the wavelength of the X-rays, $\lambda = 1.5406$ Å. After subtracting the instrumental broadening, the resultant peak broadening is related to $\beta^2_{hkl} = \beta^2_{measured} - \beta^2_{instrument}$. The instrumental broadening correction was done by using a reference standard strain-free sample. The sample was obtained by annealing the same powder at 900 °C.

Table 2 shows that the crystallite size of the β-Sn phase was reduced continuously with milling time. The final crystallite size after 45 h milling was 76 nm. The lattice strain also increased with milling time and approached a value of 1.87% after 45 h.

Table 2. Crystallite size and lattice strain.

Milling Time (h)	Crystallite Size (D, nm)	Lattice Strain (%)
5	185 ± 14	0.23 ± 0.07
15	173 ± 12	0.39 ± 0.08
25	159 ± 11	0.42 ± 0.09
35	139 ± 09	0.51 ± 0.11
45	76 ± 05	1.87 ± 0.24

The crystallite size refinement was attributed to the continuous fracturing and welding of powder particles together at the initial stages of HEBM, followed by the work hardening of powder particles [36–38]. The nanostructuring was also associated with the introduction of lattice strains due to the continuous deformation of the powder particles [26–28].

The SEM images of the Sn-0.7Cu powder after undergoing the HEBM process for various milling times are shown in Figure 2.

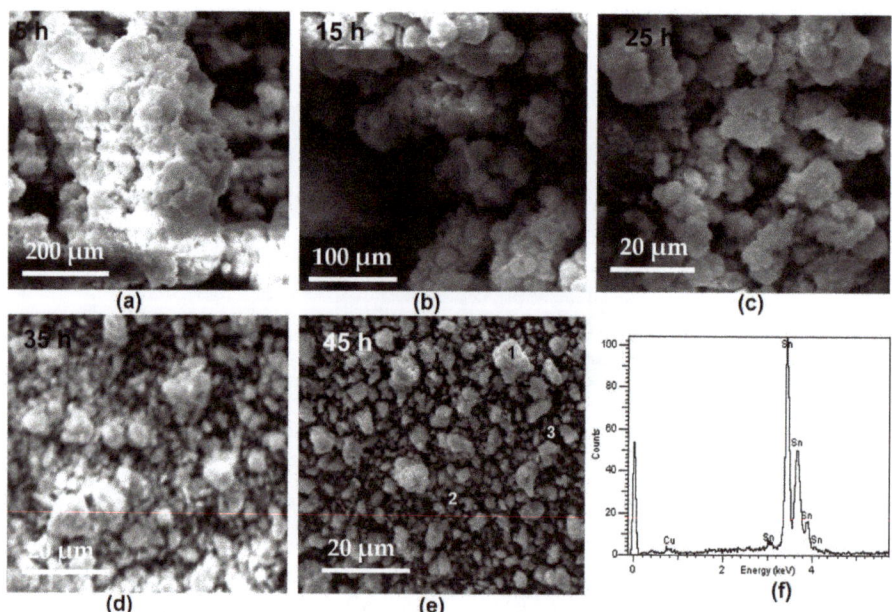

Figure 2. SEM micrograph of Sn-0.7Cu alloy milled for various times. (a) 5 h, (b) 15 h, (c) 25 h, (d) 35 h, (e) 45 h, and (f) energy-dispersive X-ray spectroscopy (EDS) analysis.

The compositional analysis was performed using the EDS analysis (Table 3). The presence of the β-Sn and Cu_6Sn_5 IMCs were noticed from Table 3 without any other impurity elements.

Table 3. Compositional analysis of Sn-0.7 wt. %Cu milled powder.

Phase	Composition (at.%)	
	Cu	Sn
Bright (1)	1.94	98.06
Dark (2)	64.7	35.3
Grey (3)	0	100

As shown in Figure 2, the particle shape was initially flaky and large fragments were present, while later stages of the milling process showed finer spherical shaped particles. In HEBM, the plastic deformation, cold welding, and refracturing of metal powders depend upon the mechanical properties of the constituent powder particles [26–28]. In the present investigation, both Cu and Sn are plastic components; therefore, they deform following the plastic-plastic particle system according to Suryanarayana [26]. These plastic–plastic couples flattened to a high surface area structure through a micro-forging process [26]. In due course, the Sn and Cu powder particles overlapped and rewelded to form a lamellar composite structure as shown in (Figure 2a,b).

A shorter diffusion path in the cold-welded lamellar structure and a higher defect ratio accelerated the alloying of Cu and Sn. The solid-state reaction between Cu and Sn led to the formation of Cu_6Sn_5 IMCs (Figure 2c). With prolonged milling times, lamellar composite fragments became work hardened and brittle [26–28]. These brittle particles fractured easily through nucleation and the growth of microcracks in big particles, which resulted in smaller spherical particles (Figure 2d). Agglomeration of these particles may also occur; however, the loose agglomerates were crushed down to fine particles (Figure 2e).

The microhardness of the samples for various milling times is shown in Figure 3. The microhardness of the Sn-0.7Cu alloy increased with milling time, reaching its maximum value at 45 h. The increase in the microhardness value was mainly related to the stored strain energy, as well as a decrease in crystallite size as already discussed. In the observed alloying process, the microhardness showed a linear dependence on the milling time.

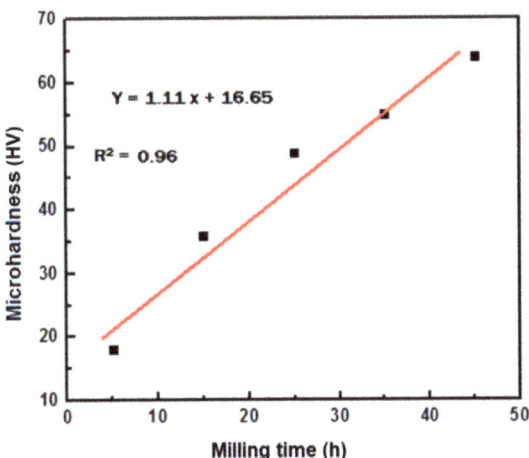

Figure 3. Vickers microhardness of the Sn-0.7Cu alloy as a function of milling time.

Accordingly, the microhardness can be linearly fitted as [39]

$$HV = U + Vt. \qquad (2)$$

Therefore,

$$\ln \frac{L_0}{L_t} = \frac{K}{V} \ln\left|1 + \frac{V}{U}t\right| \quad (3)$$

Here, K, U, and V are constants. t is the milling time, and L represents the thickness of each layer (lamellae) of composite powder. U and V indicate the intercept and slope in Figure 3. $K \approx 134$ is given by Equation (3) for $t = 5$, $L_5 = 44$ μm, and for $t = 25$ h, $L_{25} = 1.34$ μm. With continuous milling of up to 45 h, the lamellar thickness decreased severely, confirming the alloying, which is also supported by SEM results.

The thermal properties of the HEBM-ed Sn-0.7Cu alloy were assessed via DSC experiments, as shown in Figure 4.

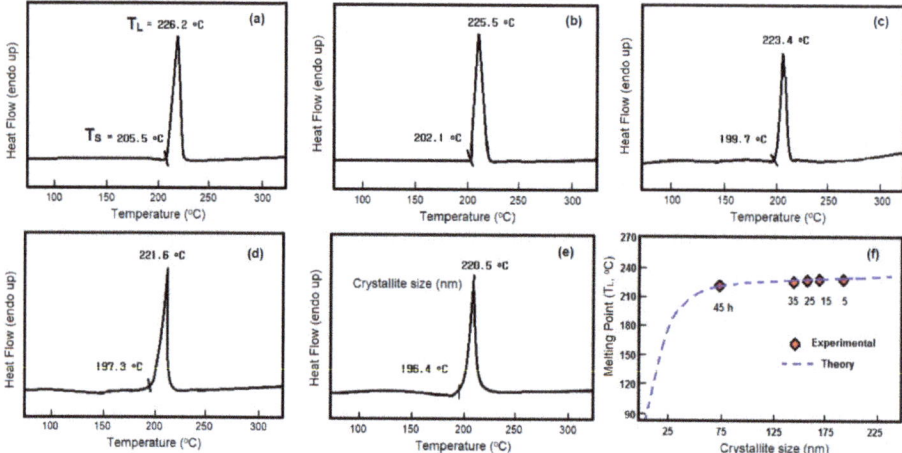

Figure 4. The differential scanning calorimetry (DSC) curve of the Sn-0.7Cu alloy after HEBM for different milling times. (a) 5 h, (b) 15 h, (c) 25 h, (d) 35 h, (e) 45 h, and (f) liquidus of the Sn-0.7Cu alloy as a function of crystallite size.

The solidus temperature (T_S) decreased down to 196.4 °C after milling for 45 h (Figure 4a–e). It is already shown that the milling induced crystal refinement and raised the stored energy, which lowered the Gibbs free energy and created a noted depression in the melting point [26,27]. The various melting temperatures (liquidus (T_L), solidus (T_S), pasty range (T_L–T_S)) and enthalpy are shown in Table 4.

Table 4. Thermal characteristics of Sn-0.7 alloy as a function of milling time.

Milling Time (h)	Solidus (T_S) (°C)	Liquidus (T_L) (°C)	Pasty Range (T_L–T_S) (°C)	Enthalpy (J/g)
5	205.5	226.2	20.7	57.8
15	202.1	225.5	23.4	57.3
25	199.7	223.4	23.7	55.4
35	197.3	221.6	24.3	53.2
45	196.4	220.5	24.1	52.1

Previous reports have shown that enthalpy changes inversely to the grain size of the powder particles [9]. This means that stored enthalpy rises as milling proceeds for a longer time. The liquidus temperature of the Sn-0.7Cu alloy was 220.5 °C after 45 h of milling, which showed an almost 7 °C depression in the melting point, compared to existing cast Sn-0.7Cu alloys in literatures [1–5]. A depression in the melting point is beneficial, as it results in the lower service temperature and increasingly cost-effective performance of electronic devices.

From Table 4, it is seen that solidus (T_S) and liquidus (T_L) temperatures shifted to a lower temperature as milling time increased. Due to the severe fracturing of the powder particles and the lattice defects induced during HEBM, the stored energy of the powder increased significantly. During heating, the deformation energy lowered the activation energy for grain growth and released the buildup of stresses in the material. A combination of recovery, recrystallization, and grain growth occurred during heating, reducing the overall stored internal energy of the system. Dislocations were eliminated/rearranged during the recovery process, followed by recrystallization and subsequent grain growth [40].

To assess the effect of milling on the melting point of Sn-0.7Cu alloy, we calculated the depression in melting point theoretically as a function of crystallite size using the following model [20,41]:

$$\Delta T = T_{m,bulk} - T_{m,r} = \frac{T_{m,bulk}}{H_{m,bulk} * \rho_s r} \left[\sigma_s - \sigma_l \left(\frac{\rho_s}{\rho_l} \right)^{\frac{2}{3}} \right] \quad (4)$$

where ΔT shows the melting point difference of bulk ($T_{m,bulk}$ = 227 °C) and nanocrystalline solid ($T_{m,r}$ < $T_{m,bulk}$), $H_{m,bulk}$ represents the enthalpy of melting of solid material ($\approx$58.6 J/g), ρ_s and ρ_l denote the density of the material in the solid and liquid phases ($\approx$7.29 g/cm^3), σ_s and σ_l denote the surface tensions of the bulk ($\approx$ 0.584 J/m^2) and liquid phase material ($\approx$0.494 J/m^2), and r is the radius of powder particles. The density and surface tension data were used from [1].

The calculated melting point data were very close to the experimentally observed melting temperatures. It is also noteworthy that the melting point decreased rapidly as soon as the particle size dropped to below $\approx$ 25 nm (Figure 4f).

The wetting of the milled powders was calculated by the area spread ratio technique. The milled Sn-0.7Cu powders were reflowed over the metallic Cu substrates at 250 °C for 70 s. The reflow profile adopted for the spread ratio measurements is shown in Figure 5.

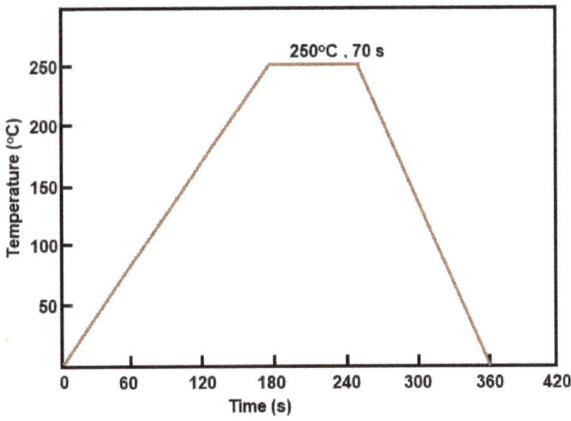

Figure 5. The reflow profile for the measurement of the solderability of the Sn-0.7Cu alloy.

The area spread ratio was calculated by the following relation:

$$\text{Spread ratio (\%)} = \frac{(A_f - A_i)}{A_i}, \quad (5)$$

Here, $A_f - A_i$ is the difference between the final and initial spread areas. The spread ratio of Sn-0.7Cu alloy as a function of milling time is shown in Table 5. The results indicate that the area spread ratio improved continuously with milling time.

Table 5. The solderability of the Sn-0.7 wt. % Cu powder after various milling times.

Milling Time (h)	Spread Ratio (%)	Error (%)
5	63	±2
15	72	±4
25	75	±5
35	80	±6
45	84	±7

It should be noted that the nanostructuring of the powder particles assisted in lowering the interfacial tension at the contact surfaces [42]. The nanostructured powder particles had higher surface energy than the unmilled powder. Therefore, the spread ratio improved [43,44]. The area spread ratio was 63% for 5 h milled powder, and increased rapidly to 80% for 35 h milled powder due to the rapid nanostructuring of powder particles. The maximum spread ratio ($\approx$84%) was achieved at a milling time of 45 h.

Spreading demonstrates the ability of the liquid metal to keep contact with the solid surface. When two contact surfaces are brought together in intimate contact, the intermolecular forces of attraction become operative. According to Thomas Young, these forces are related to the surface energy of the solid, liquid, and vapor interfaces and wetting angle given by

$$\gamma_{sv} - \gamma_{sl} = \gamma_{lv} \cos\theta \tag{6}$$

Here, γ_{sv} and γ_{lv} denote the interfacial energy of the condensed phase and γ_{sl} denotes the interfacial energy of the solid-liquid interface. Therefore, a higher spreading of the liquid indicates a smaller value of wetting angle θ.

The possible reason for the smaller wetting angle of the milled powder is that the nanostructured powder particles had a positive impact on the wettability. With an increase in milling time, the powder particles were broken down into high-surface-energy nanoparticles. The milled Sn-0.7Cu powder particles were more likely to wet the interface between the liquid Sn-0.7Cu and Cu substrate, thus lowering the interfacial surface energy between the liquid Sn-0.7Cu and Cu substrate, resulting in a smaller wetting angle and improved spreading. Therefore, the spreading of the powder increased with milling time.

For high-end electronic devices, the wettability factor is most important for soldering. A good wetting ensures the proper bonding of the solder alloy to the metallic substrate. This result depicted that the use of HEBM was advantageous to enhance the solderability of the Sn-0.7Cu alloy on copper conductors. Similarly, the thermal and microstructural properties were also improved via the HEBM approach, for potential application in Pb-free electronic devices.

4. Conclusions and Future Research Directions

1. Eutectic Sn-0.7Cu alloy was successfully produced by HEBM for 45 h. The XRD results showed that the crystallite size of the Sn-0.7Cu alloy was decreased down to 76 nm after 45 h of HEBM.
2. Microhardness of the HEBM-ed Sn-0.7Cu alloy increased continuously with milling time. The maximum hardness ($\approx$63 HV) was obtained for 45 milled powder.
3. It was shown that HEBM up to 45 h decreased the solidus and liquidus melting temperatures down to 196.4 °C and 220.5 °C, respectively.
4. There was a considerable depression of ~7 °C in the melting point of 45 h ball milled Sn-0.7Cu alloy. The spreadability of HEBM-ed Sn-0.7Cu powder improved significantly, to 84%, after 45 h of HEBM.
5. Although interesting observations have been made in this work, much work remains to study the application of the Sn-0.7Cu alloy in electronic packaging. The future work directions include the application of the Sn-0.7Cu nanopowder for the reflow soldering on Cu and electroless nickel immersion gold (ENIG) substrates in chip-scale devices.

6. The results can also be applied to study the thermomechanical fatigue and fretting-wear properties of the reflowed solders. The study of thermal stability of the developed solders at higher reflow temperatures also requires further attention.
7. Various solder alloys have already been tested for high-temperature thermal shock reliability; nevertheless, most of the Sn-Cu alloys need further research to obtain a better understanding of their performance.

Author Contributions: Conceptualization, B.A.; formal analysis, A.S. and B.A.; funding acquisition, H.S.; investigation, A.S., M.C.O., M.J.C.; methodology, A.S.; resources, H.S.; B.A.; supervision, B.A.; validation, B.A.; writing—original draft, A.S.; writing—review and editing, H.S.; B.A. All authors have read and agreed to the published version of the manuscript.

Funding: This work was supported by the Ajou University Research Fund.

Acknowledgments: This work was supported by the Ajou University Research Fund.

Conflicts of Interest: The authors declare no conflicts of interest.

References

1. Abtew, M.; Selvaduray, G. Lead-free solders in microelectronics. *Mater. Sci. Eng. R* **2000**, *27*, 95–141. [CrossRef]
2. Zeng, K.; Tu, K.N. Six cases of reliability study of Pb-free solder joints in electronic packaging technology. *Mater. Sci. Eng. R* **2002**, *38*, 55–105. [CrossRef]
3. Cheng, S.; Huang, C.M.; Pecht, M. A review of lead-free solders for electronics application. *Microelectron. Reliab.* **2017**, *75*, 77–95. [CrossRef]
4. Ma, H.; Suhling, J.C. A review of mechanical properties of lead-free solders for electronic packaging. *J. Mater. Sci.* **2009**, *44*, 1141–1158. [CrossRef]
5. Suganuma, K. Advances in lead-free electronics soldering. *Curr. Opin. Solid State. Mater.* **2001**, *5*, 55–64. [CrossRef]
6. Guo, F.J. Composite lead-free electronic solders. *Mater. Sci. Mater. Electron.* **2007**, *18*, 129–145. [CrossRef]
7. Puttlitz, K.J.; Stalter, K.A. *Handbook of Lead-Free Solder Technology for Microelectronic Assemblies*; Marcel Dekker: New York, NY, USA, 2004.
8. Kim, K.S.; Huh, H.S.; Suganuma, K. Effects of intermetallic compounds on properties of Sn–Ag–Cu lead-free soldered joints. *J. Alloy. Compd.* **2003**, *352*, 226–236. [CrossRef]
9. Wu, C.M.L.; Yu, D.Q.; Law, C.M.T.; Wang, L. Properties of Pb-free solder alloys with rare earth element additions. *Mater. Sci. Eng. R* **2004**, *44*, 1–44. [CrossRef]
10. Lee, T.Y.; Choi, W.J.; Tu, K.N.; Jang, J.W.; Kuo, S.M.; Lin, J.K.; Frear, D.R.; Zeng, K.; Kivilahti, J.K. Morphology, kinetics, and thermodynamics of solid state aging of eutectic SnPb and Pb-free solders (SnAg, SnAgCu, and SnCu) on Cu. *J. Mater. Res.* **2002**, *17*, 291–301. [CrossRef]
11. Ki, Y.S.; Kim, H.I.; Kim, J.M.; Shin, Y.E. Prediction of thermal fatigue life on mBGA solder joint using Sn-3.5Ag, Sn-3.5Ag-0.7Cu, Sn-3.5Ag-3.0In-0.5Bi solder alloys. *JWJ* **2003**, *21*, 92–98.
12. Park, J.H.; Lee, H.Y.; Jhun, J.H.; Cheon, C.S.; Jung, J.P. Characteristics of Sn–1.7Bi–0.7Cu–0.6In lead-free solder. *JWJ* **2008**, *26*, 43–48.
13. Moon, J.W.; Kim, M.I.I.; Jung, J.P. A study on the soldering characteristics of Sn–Ag–Bi–In ball in BGA. *JWJ* **2002**, *20*, 99–103.
14. Kotadia, H.R.; Mokhtari, O.; Clode, M.P.; Green, M.A.; Mannan, S.H. Intermetallic compound growth suppression at high temperature in SAC solders with Zn addition on Cu and Ni–P substrates. *J. Alloys Compd.* **2012**, *511*, 176–188. [CrossRef]
15. Myung, W.R.; Ko, M.K.; Kim, Y.J. Effects of Ag content on the reliability of LED package component with Sn-Bi-Ag solder. *J. Mater. Sci. Mater. Electron.* **2015**, *26*, 8707–8713. [CrossRef]
16. Zhang, L.; Sun, L.; Guo, Y.H. Microstructures and properties of Sn58Bi, Sn35Bi0.3Ag, Sn35Bi1.0Ag solder and solder joints. *J. Mater. Sci. Mater. Electron.* **2015**, *26*, 7629–7634. [CrossRef]
17. Zhao, M.; Zhang, L.; Liu, Z.Q.; Xiong, M.-Y.; Sun, L. Structure and properties of Sn-Cu lead-free solders in electronics packaging. *Sci. Technol. Adv. Mater.* **2019**, *20*, 421–444. [CrossRef]

18. Nishikawa, H.; Piao, J.Y.; Takemoto, T. Interfacial reaction between Sn-0.7Cu(-Ni) solder and Cu substrate. *J. Electron. Mater.* **2006**, *35*, 1127–1132. [CrossRef]
19. Kotadia, H.R.; Howes, P.D.; Mannan, S.H. A review: On the development of low melting temperature Pb-free solders. *Microelectron. Reliab.* **2014**, *54*, 1253–1273. [CrossRef]
20. Buffat, P.; Borel, J.P. Size effect on the melting temperature of gold particles. *Phys. Rev. A* **1976**, *13*, 2287–2298. [CrossRef]
21. Nanda, K.K. Size-dependent melting of nanoparticles: Hundred years of thermodynamic model. *Pramana* **2009**, *72*, 617–628. [CrossRef]
22. Letellier, P.; Mayaffre, A.; Turmine, M. Melting point depression of nanosolids: Nonextensive thermodynamics approach. *Phys. Rev. B* **2007**, *76*, 045428. [CrossRef]
23. Lee, H.Y.; Sharma, A.; Kee, S.H.; Lee, Y.W.; Moon, J.T.; Jung, J.P. Effect of aluminium additions on wettability and intermetallic compound (IMC) growth of lead free Sn (2 wt. % Ag, 5 wt. % Bi) soldered joints. *Electron. Mater. Lett.* **2014**, *10*, 997–1004. [CrossRef]
24. Shen, J.; Chan, Y.C. Research advances in nano-composite solders. *Microelectron. Reliab.* **2009**, *49*, 223–234. [CrossRef]
25. Sharma, A.; Das, S.; Das, K. Pulse Electrodeposition of Lead-Free Tin-Based Composites for Microelectronic Packaging. In *Electrodeposition of Composite Materials*; Mohamed, A.M.A., Golden, T.D., Eds.; InTech: Rijeka, Croatia, 2016; pp. 253–274.
26. Suryanarayana, C. Mechanical alloying and milling. *Prog. Mater. Sci.* **2001**, *46*, 1–184. [CrossRef]
27. Koch, C.C. Synthesis of nanostructured materials by mechanical milling: Problems and opportunities. *Nanostruct. Mater.* **1997**, *9*, 13–22. [CrossRef]
28. Koch, C.C.; Scattergood, R.O.; Youssef, K.M.; Chan, E.; Zhu, Y.T. Nanostructured materials by mechanical alloying: New results on property enhancement. *J. Mater. Sci.* **2010**, *45*, 4725–4732. [CrossRef]
29. Huang, M.L.; Wu, C.M.L.; Lai, J.K.L.; Wang, L.; Wang, F.G. Lead free solder alloys Sn-Zn and Sn-Sb prepared by mechanical alloying. *J. Mater. Sci. Mater. Electron.* **2000**, *11*, 57–65.
30. Lai, L.H.; Duh, J.G. Lead-free Sn–Ag and Sn–Ag–Bi solder powders prepared by mechanical alloying. *J. Electron. Mater.* **2003**, *32*, 215–220. [CrossRef]
31. Nai, S.M.L.; Wei, J.; Gupta, M. Lead free solder reinforced with multiwalled carbon nanotubes. *J. Electron. Mater.* **2006**, *35*, 1518–1522. [CrossRef]
32. Sharma, A.; Sohn, H.R.; Jung, J.P. Effect of graphene nanoplatelets on wetting, microstructure, and tensile characteristics of Sn-3.0 Ag-0.5 Cu (SAC) alloy. *Metall. Mater. Trans A* **2016**, *47*, 494–503. [CrossRef]
33. Babaghorbani, P.; Nai, S.M.L.; Gupta, M. Development of lead-free Sn–3.5Ag/SnO$_2$ nanocomposite solders. *J. Mater. Sci. Mater. Electron.* **2009**, *20*, 571–576. [CrossRef]
34. Gain, A.K.; Chan, Y.C.; Yung, W.K.C. Microstructure, thermal analysis and hardness of a Sn-Ag-Cu-1 wt. % nano-TiO$_2$ composite solder on flexible ball grid array substrates. *Microelectron. Reliab.* **2011**, *51*, 975–984. [CrossRef]
35. Liu, P.; Yao, P.; Liu, J. Effect of SiC nanoparticle additions on microstructure and microhardness of Sn-Ag-Cu solder alloy. *J. Electron. Mater.* **2008**, *37*, 874–879. [CrossRef]
36. Mote, V.D.; Purushuttom, Y.; Dole, B.N. Williamson-Hall analysis in estimation of lattice strain in nanometer-sized ZnO particles. *J. Theor. Appl. Phys.* **2012**, *6*, 1–8. [CrossRef]
37. Alijani, F.; Amini, R.; Ghaffari, M.; Alizadeh, M.; Okyay, A.K. Effect of milling time on the structure, micro-hardness, and thermal behavior of amorphous/nanocrystalline TiNiCu shape memory alloys developed by mechanical alloying. *Mater. Des.* **2014**, *55*, 373–380. [CrossRef]
38. Reddy, B.; Bhattacharya, P.; Singh, B.; Chattopadhyay, K. The effect of ball milling on the melting behavior of Sn–Cu–Ag eutectic alloy. *J. Mater. Sci.* **2009**, *44*, 2257–2263. [CrossRef]
39. Benjamin, J.S.; Volin, T.E. The mechanism of mechanical alloying. *Metall. Trans. A* **1974**, *5*, 1929–1934. [CrossRef]
40. Abdoli, H.; Ghanbarib, M.; Baghshahi, S. Thermal stability of nanostructured aluminum powder synthesized by high-energy milling. *Mater. Sci. Eng. A* **2011**, *528*, 6702–6707. [CrossRef]
41. Olson, E.A.; Efremov, M.Y.; Zhang, M.; Allen, L.H. Size-dependent melting of Bi nanoparticles. *J. Appl. Phys.* **2005**, *97*, 034304. [CrossRef]
42. Sharma, A.; Roh, M.H.; Jung, D.H.; Jung, J.P. Effect of ZrO$_2$ nanoparticles on the microstructure of Al-Si-Cu filler for low-temperature Al brazing applications. *Metall. Mater. Trans. A* **2016**, *47*, 510–521. [CrossRef]

43. Jung, D.H.; Sharma, A.; Jung, J.P. Influence of dual ceramic nanomaterials on the solderability and interfacial reactions between lead-free Sn-Ag-Cu and a Cu conductor. *J. Alloy Compd.* **2018**, *743*, 300–313. [CrossRef]
44. Ismail, N.; Ismail, R.; Ubaidillah, N.K.A.N.; Jalar, A.; Zain, N.M. Surface roughness and wettability of SAC/CNT lead free solder. *Mater. Sci. Forum* **2016**, *857*, 73–75. [CrossRef]

© 2020 by the authors. Licensee MDPI, Basel, Switzerland. This article is an open access article distributed under the terms and conditions of the Creative Commons Attribution (CC BY) license (http://creativecommons.org/licenses/by/4.0/).

Article

Hybrid Laser Deposition of Fe-Based Metallic Powder under Cryogenic Conditions

Aleksander Lisiecki [1,*] and Dawid Ślizak [2]

1. Department of Welding Engineering, Faculty of Mechanical Engineering, Silesian University of Technology, Konarskiego 18A Str., 44-100 Gliwice, Poland
2. Additive Manufacturing Laboratory, PROGRESJA S.A., Żelazna 9 Str., 40-851 Katowice, Poland; dslizak@progresja.co
* Correspondence: aleksander.lisiecki@polsl.pl; Tel.: +48-32-237-1649

Received: 31 December 2019; Accepted: 25 January 2020; Published: 28 January 2020

Abstract: The purpose of this study was to demonstrate the novel technique of laser deposition of Fe-based powder under cryogenic conditions provided by a liquid nitrogen bath. Comparative clad layers were produced by conventional laser cladding at free cooling conditions in ambient air and by the developed process combining laser cladding and laser gas nitriding (hybrid) under cryogenic conditions. The influence of process parameters and cooling conditions on the geometry, microstructure, and hardness profiles of the clad layers was determined. The optical microscopy (OM), scanning electron microscopy (SEM), energy-dispersive spectrometer (EDS), and XRD test methods were used to determine the microstructure and phase composition. The results indicate that the proposed technique of forced cooling the substrate in a nitrogen bath during the laser deposition of Fe-based powder is advantageous because it provides favorable geometry of the clad, low dilution, a narrow heat-affected zone, a high hardness and uniform profile on the cross-sections, homogeneity, and refinement of the microstructure. The influence of the forced cooling on microstructure refinement was quantitatively determined by measuring the secondary dendrite arm spacing (SDAS). Additionally, highly dispersed nanometric-sized (200–360 nm) precipitations of complex carbides were identified in interdendritic regions.

Keywords: laser cladding; cryogenic conditions; Fe-based coatings; fiber laser

1. Introduction

The laser beam as a heat source is widely used in different technologies of material processing and manufacturing, such surface treatment, coatings, or cutting and joining [1–8]. This is because the laser beam offers many advantages, the most important of which are high flexibility, high power densities, localized or selective heating, high processing speed, and low heat input [9–15]. One of the areas of laser beam application that is currently being widely developed is shaping the properties of surface layers and manufacturing of coatings for enhanced wear characteristics (e.g., corrosion, tribological, abrasive, thermal, and mechanical fatigue or impact load) [16–21].

Laser cladding and different methods of laser surface modification of different materials have been widely studied and much information in this area can be found in the worldwide literature. For example, an interesting and original technique of laser cladding under high frequency micro-vibration was developed and presented by Li et al. [22]. Many researchers point to several advantages of laser cladding over other methods of coatings [23–27]. However, current trends to ensure the highest energy efficiency of machines and devices, including the durability and reliability of machine components and tools, force the search for new or improved materials, as well as for new methods of their manufacturing, to overcome the limitations of current technologies. For this reason, a growing tendency to develop composite or hybrid materials and hybrid processes can be noticed [28,29]. The hybrid process combines at least two different machining processes carried out simultaneously in the same processing area (e.g., one weld pool in the case of fusion welding processes).

Among the metallic materials providing good wear resistance in various conditions, there are alloys based on Ni, Co, and Fe. However, due to the high costs, especially of Co-based alloys and additionally environmental restrictions, the interest in Fe-based alloys is increasing [30]. Sadeghimeresht et al. demonstrated that the FeCr type coatings show high corrosion resistance; thus, such coatings can be recommended as a potential alternative to Co-based coatings [30]. Wang et al. investigated the effect of V and Cr in Fe-based coatings produced by laser cladding [31]. They showed that high-quality coatings with high hardness (over 60 HRC) and wear resistance can be produced. Zhu et al. demonstrated the beneficial effect of Low-Temperature Tempering on the microstructure and properties of martensitic stainless steel coatings produced by laser cladding [32]. They proved that the post-treatment leads to martensite decomposition into finer tempered martensite with precipitations of numerous nano-sized Fe_3C carbides [32]. Chen et al. investigated the laser cladding of Fe-based coatings on rotating shafts made of 35CrMo steel, working in the seawater environment [33]. They successfully produced high-quality coatings characterized by high hardness and better corrosion resistance compared with the base metal of the substrate [33].

Another interesting example of manufacturing a high performance Fe-based coating is the study conducted by Li et al. [34], which was an investigation of the effect of carbon fibers addition to Fe-based coatings produced by laser cladding. They demonstrated a significant increase in the microhardness and wear resistance of the composite coatings containing also nano-size carbides [34].

Hou et al. investigated the influence of scanning speed during the laser cladding of Fe-based amorphous coatings on the microstructure and properties of coatings [35]. They demonstrated that the laser cladding process could be successfully applied for manufacturing high-quality amorphous coatings. They also demonstrated that the low cost of Fe-based amorphous alloys is an advantage [35]. Zhang et al. demonstrated that deep cryogenic treatment, applied after laser cladding Fe-based coatings, can significantly enhance the microhardness and wear resistance of such coatings [36].

An original technique of laser surface melting of non-ferrous alloys in liquid nitrogen was elaborated and demonstrated by Zieliński et al. [37–39] and Cui et al. [40]. All these researchers point to the benefits of accelerated cooling and solidification rates due to liquid nitrogen-assisted cooling, which results in refinement of the microstructure. For example, Zieliński et al. reported that the laser surface melting of titanium alloy at cryogenic conditions resulted in obtaining a very hard surface layer composed of titanium nitrides and martensite structure saturated with nitrogen [37].

Cui et al. conducted laser surface melting of magnesium alloy also at cryogenic conditions provided by a liquid nitrogen bath. Cui et al. indicated that thanks to the rapid quenching associated with the process, the microstructure of the treated magnesium alloy was refined. Moreover, nanocrystals or even amorphous structures can be obtained under such conditions [40]. They also reported that the corrosion resistance of the surface layer remelted under cryogenic conditions was enhanced significantly [40].

Assuming a favorable effect of accelerated cooling also during laser powder cladding, an attempt has been made to conduct preliminary tests related to the laser deposition of Fe-based powder on a steel substrate immersed in a liquid nitrogen bath. Due to the presence of evaporated gaseous nitrogen in the region of deposition and melt pool, the novel process was considered as a hybrid process combining laser powder cladding and laser gas nitriding.

The level of difficulty of laser powder cladding under cryogenic conditions is much higher than the laser surface melting of the substrate immersed in a liquid nitrogen bath.

According to the authors' knowledge and experience, the presented results are unique, original, and not yet available in the literature.

2. Materials and Methods

The substrate for laser cladding tests was non-alloy structural steel S235JR (according to EN 10025-2) with dimensions of 100 mm × 100 mm and 5 mm thick (Table 1, Figure 1). Such non-alloy steel was chosen to minimize the effect of the substrate material on the course of cladding process, as well as on the composition of the clad layers, at the preliminary stage of the study.

Table 1. Chemical composition of non-alloy structural steel S235JR (EN 10025-2) according to the Thyssenkrupp manufacturer's data (wt. %), Figure 1.

C	Mn	Si	P	S	N	Cu	Al	Fe
0.05–0.14	0.2–0.8	0.1	0.025	0.015	0.01	0.2	0.015–0.08	Bal

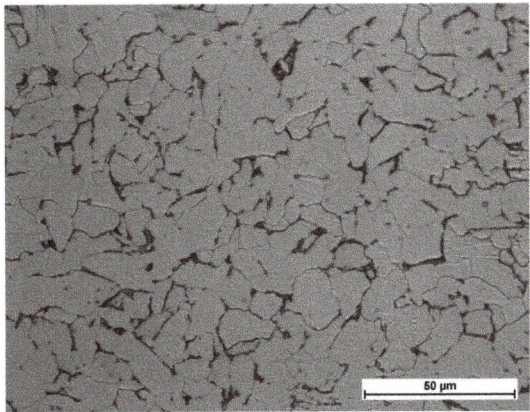

Figure 1. Microstructure of the substrate of non-alloy steel S235JR (Magnification 500×), Table 1.

The composition of the metallic powder (not commercially available) for laser cladding was designed to provide high abrasive wear resistance and good corrosion resistance. Intentionally, the amount of strong carbide-forming elements such Cr, W, and V was high enough to ensure the in situ precipitation of fine dispersive carbides during laser cladding, as shown in Table 2. The particle size distribution of the metallic Fe-based powder was 53–150 µm.

Table 2. Nominal composition of Fe-based experimental powder (wt. %).

Mo	Cr	W	V	C	Fe
8.0	5.0	2.5	2.5	1.1	Bal

In order to conduct the technological tests of cladding at cryogenic conditions, a prototype stand was designed and custom made, as shown in Figure 2a. The powder was dosed by a disc rotary feeder PFU4 (Durum, Willich, Germany) in argon carrying gas at the pressure of 1.5 bar and the rate flow of 8.0 L/min. Powder was gravity transported from a disc feeder to the nozzles head through flexible PTFE tubes (Polytetrafluoroethylene—"Teflon") with an inner diameter of 2.0 mm each. Three individual and cylindrical nozzles of 0.8 mm diameter for powder injection were mounted on the head, as shown in Figure 2a. The powder injection nozzles were adjustable and set in such a way to focus the three individual powder streams on the top surface of the melt pool in the region of laser beam interaction, as shown in Figure 2b. The nozzles had was integrated with the laser focusing head coupled by a fiberglass with the IPG YLS laser generator (IPG Photonics, Oxford, MA, USA), emitting at the wavelength of 1.07 µm. The positioning of the laser focusing head with coaxial powder injection nozzles was provided by a robot Panasonic GII TL-190 (Industrial Solutions Company, Panasonic Corporation, Osaka, Japan). The beam parameter product (BPP) of the laser beam is <4.0 mm·mrad, and it is characterized by a Gaussian energy distribution across the beam spot (Transverse Electromagnetic Mode; TEM_{00}). With the applied configuration of the optics, the nominal beam spot diameter was 300 μm, which is typical for laser welding. In the case of laser cladding, a wider beam is advantageous; therefore, the laser beam was defocused by lifting the laser head. Thus, the focal plane was 30.0 mm over the top surface of the substrate (focal spot position +30 mm), which resulted in increasing the diameter of the region of laser beam irradiation on the substrate up to 1.3 mm. Prior to cladding tests, the steel substrate was sandblasted, providing repeatable surface conditions with the roughness of Ra 25–60 μm, after which it was cleaned by ethanol C_2H_5OH.

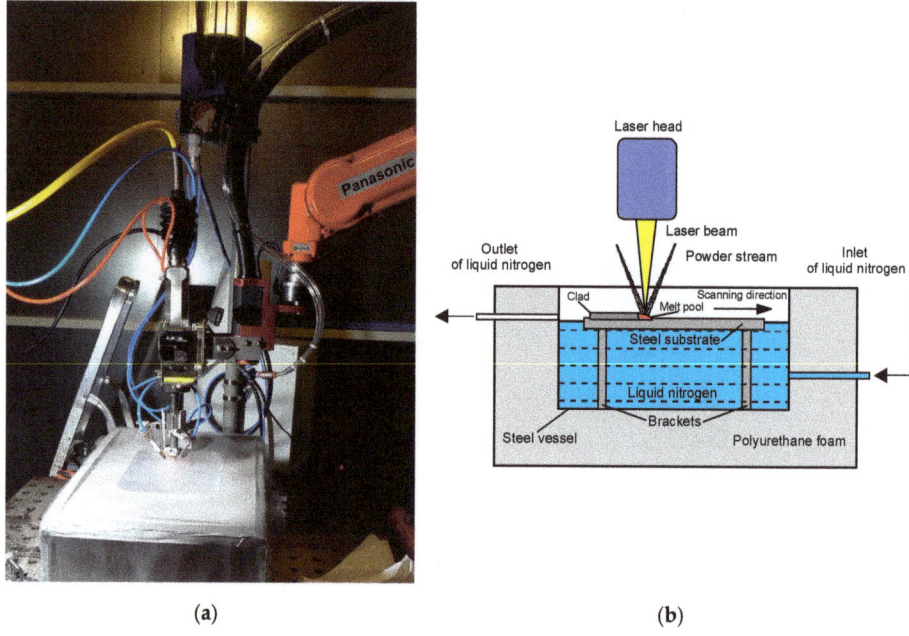

Figure 2. (**a**) A view of the laser powder deposition process under cryogenic conditions on the experimental stand for technological tests of laser cladding and powder deposition equipped with a six-axis robot, an high power fiber laser (HPFL), powder dosing, and a delivery system with a specially designed powder head with three individual adjustable nozzles and a bath of liquid nitrogen for cooling the substrate; (**b**) a scheme for the laser powder deposition of a steel specimen immersed in a liquid nitrogen bath.

The stringer beads were 40.0 mm long, while the shift between individual beads was 10.0 mm. The range of the processing parameters was determined during preliminary tests, and the laser out power ranged between 250 and 1000 W, while the scanning speed ranged between 250 and 1000 mm/min, (Table 3, Figure 3). The powder feeding rate was kept constant at 6.0 g/min. Sixteen stringer beads were produced at different processing parameters on each substrate specimen, as shown in Figure 3. One specimen was laser cladded at free cooling in ambient air at the temperature of approximately 22 °C (conventional laser cladding). Meanwhile, the second specimen was partially immersed (approximately half the specimen thickness) in the bath of liquid nitrogen at the temperature of −190 °C to provide cryogenic conditions for cooling the substrate, as shown in Figure 2a,b. The liquid nitrogen bath was thermally isolated by polyurethane foam 15.0 mm thick placed on the bottom and side walls. As can be seen in Figure 2a, the liquid nitrogen was evaporating intensively during cladding; therefore, the gaseous nitrogen was present also in the region of powder deposition. Since the nitrogen is active gas, it acts as an alloying medium during laser powder deposition. Considering the above, the process of laser deposition of the powder can be defined as a hybrid process, combining conventional laser cladding and alloying or gas nitriding, which was additionally conducted under cryogenic conditions.

Table 3. Parameters of laser deposition of experimental Fe-based metallic powder at free cooling conditions (conventional laser cladding) and forced cooling by liquid nitrogen bath under cryogenic conditions (hybrid laser deposition process), as shown in Figure 3.

No.	Surface Layer Indication Free Cooling/Cryogenic	Scanning Speed (mm/min)	Laser Power (W)	Energy Input (J/mm)	Remarks Free Cooling/Cryogenic
1	LC1/HC1	250	250	60	HQ/HQ
2	LC2/HC2	250	500	120	UB/HQ, SP
3	LC3/HC3	250	750	180	UB, SP/UB
4	LC4/HC4	250	1000	240	HQ, SP/UB, SP
5	LC5/HC5	500	250	30	HQ/HQ
6	LC6/HC6	500	500	60	HQ/UB
7	LC7/HC7	500	750	90	HQ, SP/UB
8	LC8/HC8	500	1000	120	UB, V/HQ
9	LC9/HC9	750	250	20	HQ/LF
10	LC10/HC10	750	500	40	HQ, SP/IF
11	LC11/HC11	750	750	60	UB/IF
12	LC12/HC12	750	1000	80	HQ/UB
13	LC13/HC13	1000	250	15	IF/LF
14	LC14/HC14	1000	500	30	HQ/LF
15	LC15/HC15	1000	750	45	UB/IF
16	LC16/HC16	1000	1000	60	UB/UB

Remarks: UB—uneven bead, SP—single pore, IF—incomplete fusion, V—voids, HQ—high quality, LF—lack of fusion.

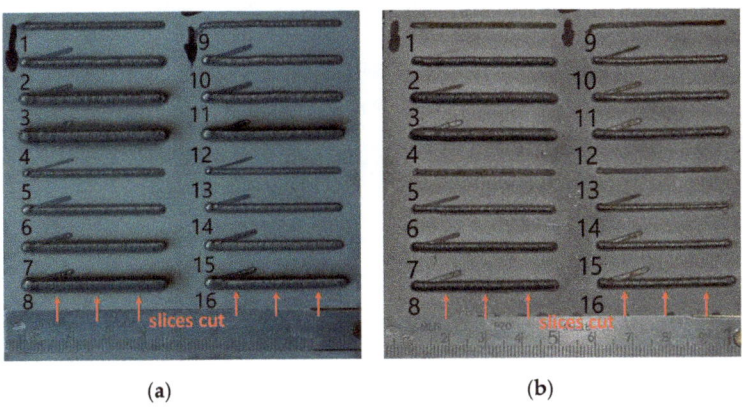

Figure 3. A view of stringer beads produced by the laser deposition of experimental Fe-based metallic powder at (a) free cooling conditions (conventional laser cladding); (b) and forced cooling by liquid nitrogen bath under cryogenic conditions (hybrid laser deposition process), as shown in Table 3.

After completing the laser cladding tests, the first visual inspection was done. Next, the cross-sections were cut, and samples for metallographic study were prepared. Three sliced sections for every individual tested clad layer were take—one from the middle region of a bead, and two other at the distance 10.0 mm from the ends, as shown in Figure 3. A standard metallographic procedure for preparing the samples was applied.

The sections were mounted in thermosetting phenol resin with graphite filler Electro-WEM (Metalogis, Warsaw, Poland), next wet grinded by papers with the grit 120 to 2500, polished with 1 μm diamond suspension Metkon Diapat-M (Metkon Instruments Inc., Bursa, Turkey), and finally etched by the HNO₃ + 3HCL reagent. Metallographic tests and analysis were carried out by means of stereoscopic microscope OLYMPUS SZX9 (Olumpus Corporation, Tokyo, Japan), the inverted metallographic microscope NIKON Eclipse MA100 (Nikon Corporation, Tokyo, Japan), and scanning electron microscope (SEM) (Carl Zeiss, Oberkochen, Germany), which were equipped with the energy-dispersive spectrometer (EDS) (Oxford Instruments, Abingdon, UK) and X-Ray diffractometer (Panalitycal, Almelo, The Netherlands), with CuKα source of radiation, and the 2θ angle range of 0 to 140°.

Hardness measurements were carried out on the cross-sections along the axis of symmetry of the individual bead, starting from under the top surface region (presented also in Figure 10). The measurements were carried out by means of the hardness tester WILSON WOLPERT 401 MVD (Wolpert Wilson Instruments, Aachen, Germany) at the load of 10 N and the dwell time of 10 s. The distance between the subsequent points was 0.3 mm.

3. Results and Discussion

3.1. Macro Observations and Single Bead Geometry

In order to determine the effect of the proposed novel technique of laser cladding of Fe-based powder combined with nitriding and forced cooling the substrate at cryogenic conditions, the comparative study of the clad layers produced at conventional laser cladding, and also at the so-called "hybrid" cryogenic laser deposition, are presented below. The geometry of a single bead was determined on cross-sections by measuring the penetration depth, width, and height of a bead; the area of the fusion zone; and the total area of the clad, as shown in Figure 4. All the tested clad layers were free of cracks; however, single pores with a small diameter below 0.1 mm can be observed on the cross-sections of layers produced both during free cooling and cryogenic conditions. Thus, no relationship was found between the conditions of cooling and tendency to porosity, as shown in Figures 5–8.

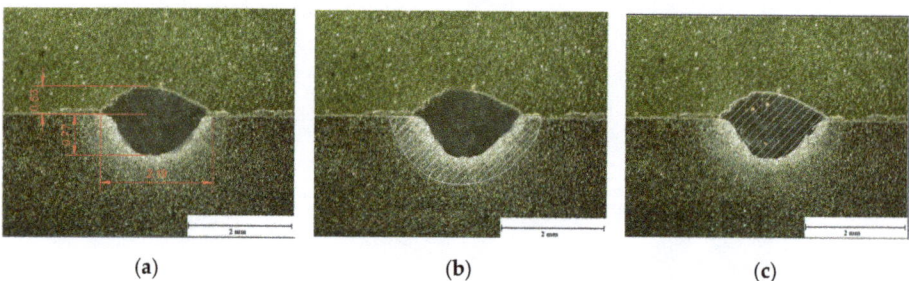

Figure 4. An example of the measuring procedure for determining the characteristic geometry features and dimensions of the stringer beads (Table 4): (**a**) measuring of the bead dimensions; (**b**) determination of the heat-affected zone (HAZ) area; (**c**) determination of the total area of the clad layer (fusion zone + clad layer).

Table 4. Results of geometry determined for the stringer beads produced by the laser deposition of experimental Fe-based metallic powder at free cooling conditions (conventional laser cladding) and forced cooling by liquid nitrogen bath under cryogenic conditions (hybrid laser deposition process), as shown in Figures 5–8.

Surface Layer	Energy Input (J/mm)	Penetration Depth (mm)	Height of Clad (mm)	Width of Clad (mm)	Total Area of Clad (mm^2)	Area of Fusion Zone (mm^2)	Dilution (%)
LC1	60	0.08 ± 0.0016	0.17 ± 0.0014	1.35 ± 0.074	0.0901	0.0324	36
LC2	120	0.55 ± 0.009	0.25 ± 0.0038	1.89 ± 0.0154	0.8610	0.5762	67
LC3	180	0.82 ± 0.011	0.42 ± 0.0067	2.22 ± 0.0167	1.7102	1.2314	72
LC4	240	0.91 ± 0.013	0.85 ± 0.014	2.57 ± 0.081	2.9313	1.4363	49
LC5	30	0.11 ± 0.0019	0.14 ± 0.0009	1.17 ± 0.068	0.1921	0.0653	34
LC6	60	0.42 ± 0.007	0.22 ± 0.0091	1.71 ± 0.0148	0.7108	0.4691	66
LC7	90	0.56 ± 0.0084	0.33 ± 0.0037	2.03 ± 0.0151	1.114	0.7130	64
LC8	120	0.77 ± 0.01	0.53 ± 0.0088	2.19 ± 0.017	1.7404	0.9310	54
LC9	20	0.07 ± 0.0012	0.05 ± 0.0008	0.69 ± 0.0113	0.0601	0.0288	48
LC10	40	0.34 ± 0.004	0.08 ± 0.0013	1.42 ± 0.079	0.3611	0.2817	78
LC11	60	0.5 ± 0.0081	0.13 ± 0.0011	1.64 ± 0.081	0.6208	0.5090	82
LC12	80	0.67 ± 0.0093	0.24 ± 0.0038	1.76 ± 0.013	1.0105	0.7680	76
LC13	15	0.04 ± 0.0006	0.04 ± 0.0006	0.58 ± 0.0093	0.0312	0.0150	48
LC14	30	0.26 ± 0.004	0.102 ± 0.0021	1.28 ± 0.073	0.2704	0.2055	76
LC15	45	0.42 ± 0.0067	0.13 ± 0.0009	1.54 ± 0.084	0.5312	0.4409	83
LC16	60	0.57 ± 0.0078	0.24 ± 0.003	1.70 ± 0.0138	0.8204	0.5661	69
HC2	120	0.05 ± 0.0062	0.55 ± 0.0012	0.62 ± 0.0096	0.8609	0.0086	1.6
HC3	180	0.15 ± 0.0067	0.80 ± 0.0096	0.90 ± 0.0128	1.7113	0.0171	3.6
HC4	240	0.21 ± 0.0083	1.25 ± 0.071	1.10 ± 0.064	2.9307	0.0293	5.2
HC6	60	0.05 ± 0.0007	0.31 ± 0.0039	0.49 ± 0.0076	0.7112	0.0071	1.1
HC7	90	0.09 ± 0.0012	0.52 ± 0.0081	0.75 ± 0.014	1.105	0.0110	1.2
HC8	120	0.16 ± 0.0012	0.91 ± 0.0121	0.82 ± 0.0119	1.7408	0.0174	2.8

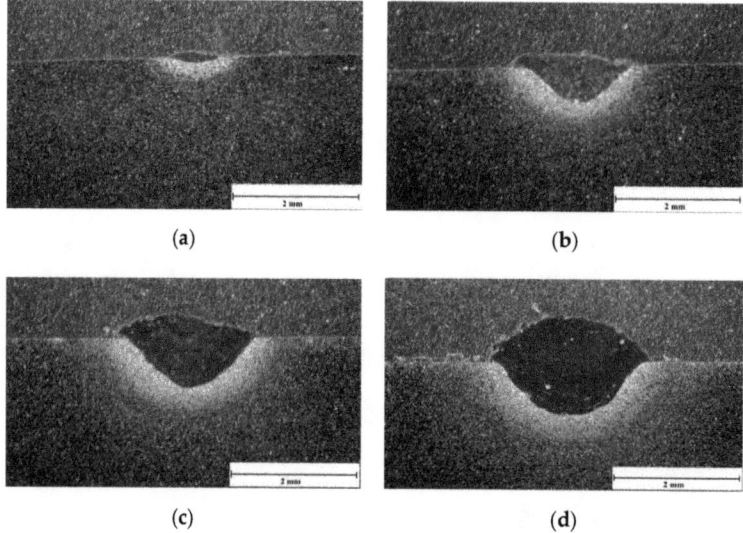

Figure 5. Macrostructure and the single bead geometry of the clad layers produced by laser cladding of Fe-based powder on steel substrate at free cooling, constant scanning speed 250 mm/min, and different laser output power, Tables 3 and 4: (**a**) LC1, 250 W; (**b**) LC2, 500 W; (**c**) LC3, 750 W; (**d**) LC4, 1000 W.

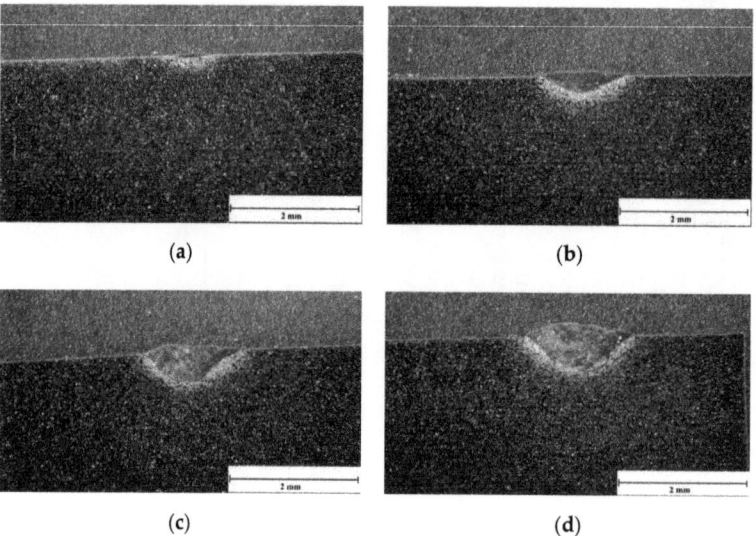

Figure 6. Macrostructure and the single bead geometry of the clad layers produced by laser cladding of Fe-based powder on steel substrate at free cooling, constant scanning speed 1000 mm/min, and different laser output power, as shown in Tables 3 and 4: (**a**) LC13, 250 W; (**b**) LC14, 500 W; (**c**) LC15, 750 W; (**d**) LC16, 1000 W.

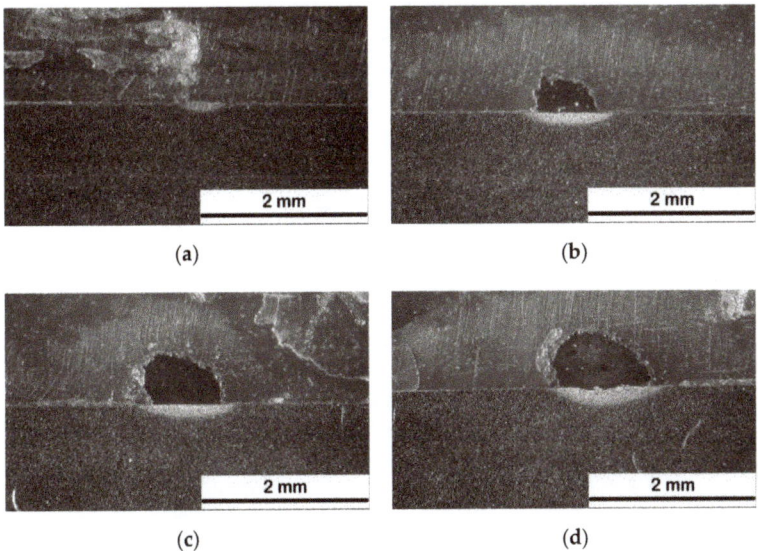

Figure 7. Macrostructure and the single bead geometry of the clad layers produced by laser deposition of Fe-based powder on steel substrate at forced cooling by a liquid nitrogen bath under cryogenic conditions (hybrid laser deposition process), a constant scanning speed of 250 mm/min, and different laser output power, as shown in Tables 3 and 4: (**a**) HC1, 250 W; (**b**) HC2, 500 W; (**c**) HC3, 750 W; (**d**) HC4, 1000 W.

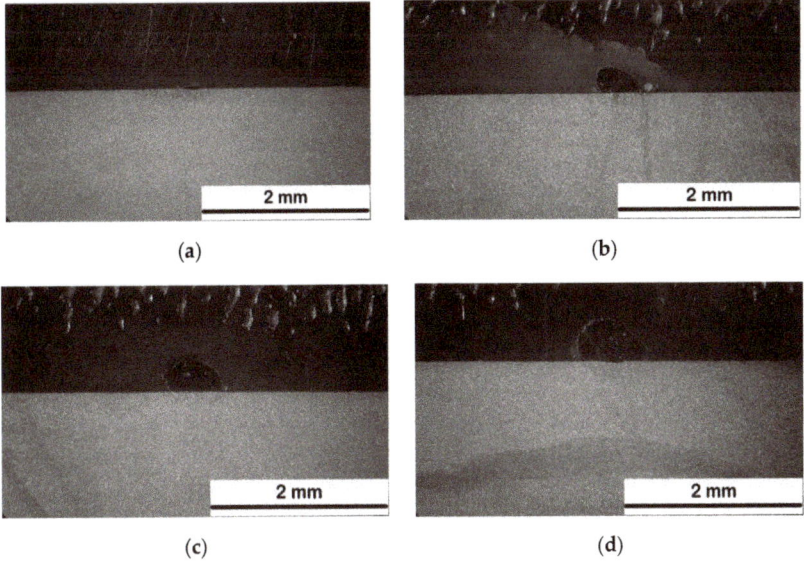

Figure 8. Macrostructure and the single bead geometry of the clad layers produced by laser deposition of Fe-based powder on steel substrate at forced cooling by liquid nitrogen bath under cryogenic conditions (hybrid laser deposition process), constant scanning speed 500 mm/min, and different laser output power, as shown in Tables 3 and 4: (**a**) HC5, 250 W; (**b**) HC6, 500 W; (**c**) HC7, 750 W; (**d**) HC8, 1000 W.

The representative cross-sections of test clad layers produced by conventional laser cladding at free cooling conditions are presented in Figures 5 and 6, while the clad layers produced under cryogenic conditions are presented in Figures 7 and 8. It is worth noting that three cross-sections were analyzed for every individual bead. Since the total number of analyzed cross-sections is huge, just the selected representative sections are presented in the following figures. The results of geometry measurements for induvial beads are summarized in Table 4, Figure 9. The given results are the mean values with standard deviation taken from the analysis of three individual sections for every single bead. The value of dilution "D" was calculated by the following formula:

$$D = \frac{A_{FZ}}{A_{FZ} + A_{CL}} \cdot 100\%, \qquad (1)$$

where A_{FZ} is the cross-section area of the fusion zone, and the A_{CL} is the cross-section area of the clad layer (presented also in Figure 10).

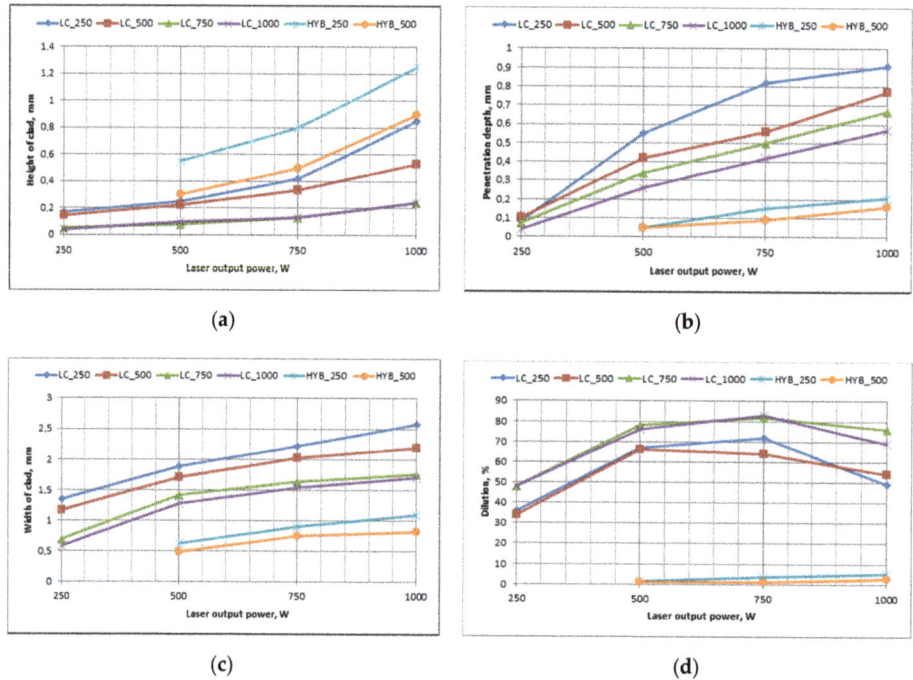

Figure 9. Influence of the process parameters and cooling conditions on the geometry and dimensions of the stringer beads produced by laser deposition of Fe-based powder (Tables 3 and 4): (a) height of the clad (reinforcement); (b) penetration depth into the substrate; (c) width of the single clad layer; and (d) calculated rate of dilution.

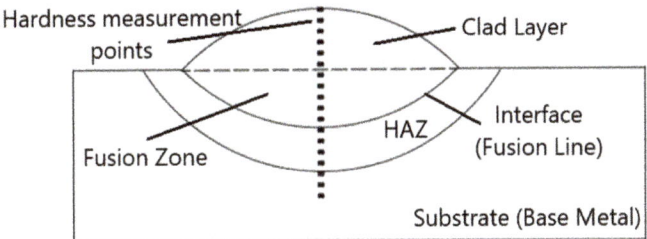

Figure 10. Scheme of hardness measurements and characteristic features of single bead geometry, as shown in Figure 11.

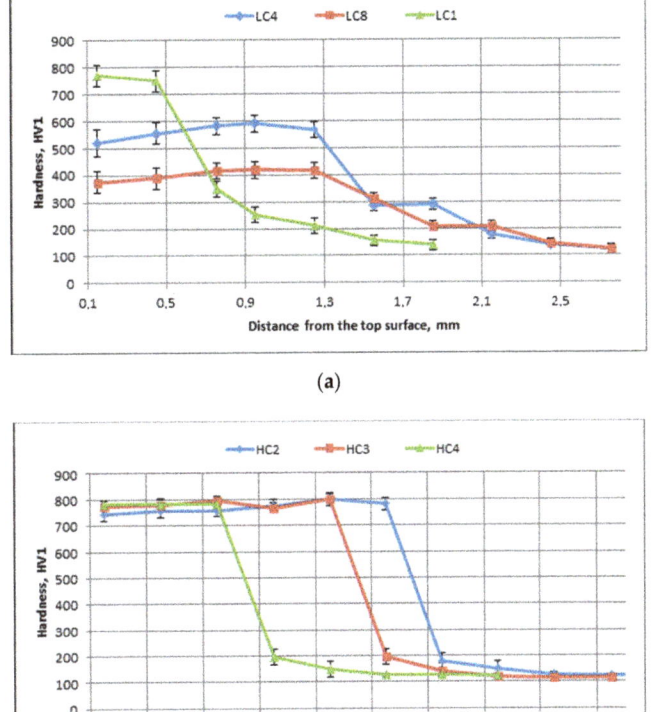

Figure 11. Hardness distribution on cross-sections of the representative stringer beads produced by the laser deposition of experimental Fe-based metallic powder (Table 3, Figure 10): (**a**) at free cooling conditions (conventional laser cladding); and (**b**) at forced cooling by liquid nitrogen bath under cryogenic conditions (hybrid laser deposition process).

The relationship between process parameters, cooling conditions, and the geometry of single beads is summarized in Figure 9. Due to the large number of curves presented on the individual graphs, and relatively low standard deviation, it was decided to provide the statistical results just in tabular form in Table 4. Presenting the standard deviation for each point on the graphs would make it completely illegible.

As can be seen, the clad layers produced by laser cladding at free cooling conditions are characterized by a clear fusion line (also called the "interface line"), indicating complete fusion across the layer, and proper metallurgical bonding in a whole range of processing parameters. It is also obvious that the shape and dimensions of the single bead depends on processing parameters; however, the penetration depth is relatively high compared to the height of the test beads, resulting in high dilution ranged from approximately 34% to over 80%, as shown in Table 4 and Figures 5, 6 and 9. Contrary, in the case of the test clad layers produced at cryogenic conditions, the geometry is completely different, as shown in Figures 7 and 8. The clad layers produced in the lowest range of energy input at a scanning speed of over 500 mm/min were characterized by incomplete fusion along the stringer beads, which was revealed during preparation of the metallographic samples. Therefore, the layers were not taken into further study.

As can be seen in Figures 7 and 8, the penetration depth is typically very small on the verge of lack of fusion. Low penetration depth resulted in very low dilution ranging from 1.1% to 5.2%, as shown in Table 4 and Figure 9. Low dilution is beneficial because it provides the chemical composition of the deposited layer similar to the original composition of the powder. Moreover, the analyzed single beads showed greater height, at the same time with a smaller width, compared to the beads produced at the same processing parameters but under free cooling conditions.

This phenomenon, as well as the low penetration depth, is related with the supercooled substrate immersed in the bath of liquid nitrogen at temperature—190 °C. Under such conditions, the supercooled substrate dissipates the absorbed laser energy, limiting the penetration ability of the laser beam. At the same time, powder particles fed into the region of laser beam interaction are heated, melted, and deposited on the substrate. Moreover, conditions for wetting the slightly melted substrate at obviously lower temperature are not favorable, resulting in trends for shaping a spherical cross-section of the single bead, as shown in Figures 7 and 8. Therefore, the single beads produced at cryogenic conditions are characterized by small width and relatively large height compared to the beads produced at free cooling conditions, as shown in Figure 9.

The above findings indicate that a controllable forced cooling of the substrate during laser cladding can be considered as an auxiliary technique for shaping the geometry of clad layers and minimizing the dilution of the deposit by the substrate material. Providing low dilution is especially of importance during the manufacturing of single layer, thin coatings on the substrate that significantly differ in chemical composition from the coating material.

3.2. Microhardness

The profiles of hardness were determined on a cross-section of the chosen representative clad layers as demonstrated in Figure 10. The profiles of Vickers HV1 hardness presented in Figure 11 showed distinct difference in the measured maximum hardness values for the clads produced at free cooling conditions and the clads produced by the novel technique of laser deposition of powder under cryogenic conditions in a liquid nitrogen bath. Additionally, the shape (distribution) of the hardness HV1 profiles is different, as can be seen in Figure 11. The presented results of hardness measurement for each point on the graphs are a mean value taken from three different sections for every tested clad layer (single bead). As can be seen, the low values of standard deviation indicate a small dispersion of results. This is due to the relatively large size of the Vickers's indenter imprint at the load of 10N, but also the homogeneity of the macrostructure, as can be seen in Figures 5–8.

In a case of the tested clad layers produced by conventional laser cladding at free cooling, the hardness reaches the maximum in the clad region, but the spread of the values differs distinctly for the individual clads (750–780 HV1 for LC1, 520–590 HV1 for LC4, and 370–420 HV1 for LC8), indicating the significant influence of processing parameters on the hardness, as shown in Figure 11a. The differences in the hardness values of the individual layers are related to the dilution degree. As can be seen in

Figure 11a, the maximum hardness value (780HV1) was measured for the clad layer characterized by the lowest dilution 36%, as shown in Figure 9d and Table 4. In turn, the higher the dilution degree (49% for LC4 and 54% for LC8), the lower the hardness in the clad region, as shown in Figure 11a and Table 4. Since the final composition of the clad layer is the effect of mixing the deposited powder and melted substrate, the increasing penetration depth leads to deterioration of the originally designed composition of the powder. In this study, the clad layers are diluted mainly by Fe, which comes from the non-alloy steel substrate, as shown in Tables 1 and 4.

At a certain depth related to the clad thickness (both the clad height and penetration depth), the hardness decreases gradually until reaching the value of the base metal (S235JR) of the substrate at approximately 120–130 HV1. The gradual decrease in hardness occurs in the heat-affected zone (HAZ). Based on the hardness distribution in the HAZ, the width of this region can be specified at 0.8 to 1.3 mm for the tested clads. It can be also verified on the cross-sections in Figures 5 and 6. Since the fusion line is considered as an interface line for a coating, and the HAZ is considered as the interface region, it can be concluded that in the case of clad layers produced by conventional laser cladding at free cooling, the interface region is "wide and smooth".

On the other hand, the profiles of hardness determined on the cross-sections of the clad layers produced under cryogenic conditions are almost flat in the clad region, while the maximum hardness value for the individual layers is in the same range between 740 and 800 HV1, as shown in Figure 11b. The hardness profile drops down sharply to approximately 200 HV1 when reaching the fusion line (interface line). As can be seen in Figure 11b, the region of the HAZ is very narrow, because the hardness of the base metal (S235JR) is reached after less than 0.5 mm. So, in this case, it can be concluded that the interface region of clad layers produced under cryogenic conditions is "narrow and sharp". This phenomenon is a result of extremely low dilution below 5% and also a very narrow HAZ due to initial super cooling the substrate in the bath of liquid nitrogen, as well as rapid heat dissipation during the laser deposition of the powder.

3.3. Microstructure and Phase Composition

The X-ray diffraction patterns of two comparative samples and the steel substrate are presented in Figure 12. The comparative samples were prepared especially for XRD analysis by multi-bead cladding of the substrate with an overlap of approximately 25% to produce the coating 6.0–7.0 mm wide for providing reliable conditions for the analysis, since the beam diameter in the applied apparatus is nominally 4 mm and can be additionally focused to 1–2 mm. The parameters providing the widest single beads produced by laser cladding both at free cooling and under cryogenic conditions were selected (CL4 and HC4, Table 3), as shown in Figure 13. Next, the surfaces were grinded to provide roughness Ra of 0.8–1.1 μm.

The X-ray diffraction pattern of the substrate ("substrate" red line in Figure 12) indicated dominant peaks from Fe-α, which is typical for non-alloy steel. In the case of the X-ray diffraction patterns of the tested coatings, the dominant peaks come from Fe-α and Fe-γ. However, clear differences can be observed in the patterns. Intensity of the Fe-α peaks is significantly higher in the case of the coating produced by laser cladding at free cooling conditions, indicating a higher share of the Fe-α in this case ("FeCr_3" green line in Figure 12). It is related both with the different degree of dilution and different cooling and solidification rates. The degree of dilution for the clad layer HC4 produced under cryogenic conditions is approximately just 5% and almost 10 times higher (approximately 49%) in the case of the clad layer LC4 produced by conventional laser cladding at free cooling conditions, as shown in Table 4. Therefore, the coating produced under cryogenic conditions is slightly influenced by the composition of the substrate of non-alloy steel, while the coating produced at free cooling is composed just of half of the powder and half of the base metal substrate. However, what is decisive for the share of Fe-α and Fe-γ is the cooling rate, which is significantly higher in the case of cryogenic conditions. An extremely high cooling rate inhibits the martensitic transformation. Therefore, the share of retained austenite is higher in the clad layer produced under cryogenic conditions ("FeCr_1" blue line in Figure 12), despite the high content of ferrite-stabilizing elements such Mo, V, W, and Cr, as shown in Table 2. Other peaks from the X-ray diffraction patterns correspond mainly with carbides $M_{23}C_6$ and M_7C_3, as shown in Figure 12. The detection level of the applied XRD method is about 3% for the specific phase; therefore, it doesn't allow for precise identification of the fine precipitations at a low share. However, due to the composition containing W, V, and Cr, complex carbides are most expected.

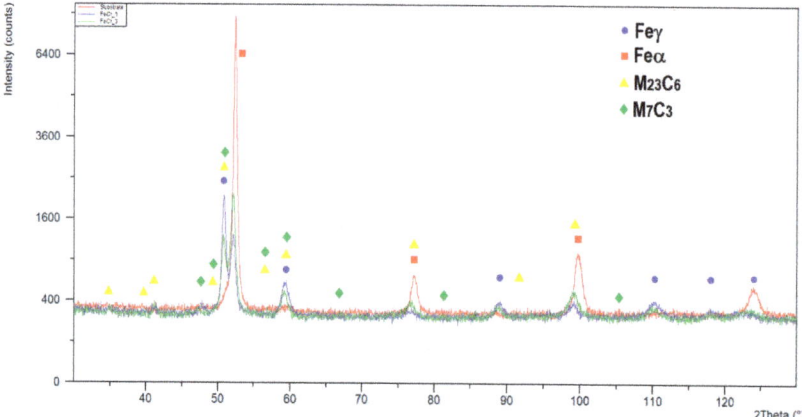

Figure 12. XRD patterns of the substrate of non-alloy steel S235JR and comparative coatings produced as multi-bead cladding at the processing parameters for LC4 and HC4 (scanning speed 250 mm/min, laser output power 1000 W, energy input 240 J/mm, free cooling and forced cooling respectively, Table 3).

Figure 13. A view of specimens prepared for XRD analysis (from left: HC4 and CL4; see Table 3).

Typical microstructures of the clad layers produced at a scanning speed of 250 mm/min and laser output power of 1000 W both at free cooling and under cryogenic conditions are presented on optical micrographs in Figure 14. Figure 14a,b show the overall microstructure in the fusion line (interface) region at the same magnification: 200×. Meanwhile, Figure 14c,d show the typical microstructure in the middle region of the clad layers at the same magnification: 500×. As can be seen, there is a clear difference in grain size. The substrate of non-alloy steel S235JR after hot rolling originally exhibits a ferritic microstructure with pearlitic regions along the grain boundaries, as shown in Figure 1. In the case of the analyzed surface layer (LC4) produced at free cooling, the region adjacent to the fusion line (interface) exhibits a microstructure of upper bainite, as shown in Figure 14b. In turn, the same region in the case of the surface layer produced under cryogenic conditions (HC4) can be identified as martensite microstructure, as shown in Figure 14a. The micrographs of the representative clad layers are free of cracks or pores. Close observation of the fusion zone (clad region) revealed the layer of planar crystallisation with a thickness of approximately 5 µm in the case of the clad layer produced at free cooling (LC4), which was three times narrower in the case of the layer produced under cryogenic conditions (HC4), as shown in Figure 14a,b. Above the planar crystals zone, the columnar dendrites can be observed with interdendritic regions. Next, in the middle region of the clad layers (fusion zone), the presence of equiaxed dendrites and cellular crystals was confirmed, as shown in Figure 14c,d. Such columnar dendrites adjacent to the interface indicate a very high G/R ratio in this region during solidification, where G is the temperature gradient, and R is the solidification rate. As the distance from the interface (fusion line) increases, the G/R ratio decreases, and equiaxed dendrites growth take place. On the other hand, the solidification rate affects the grain size, and it is related with the secondary dendrite arm spacing (SDAS). Based on the optical micrographs, the mean SDAS for the middle region of the clad layers was calculated by dividing the measuring length on the columnar dendrite by the number of dendrite arms, as shown in Table 5. The mean SDAS for the clad layer produced at free cooling was 3.1 µm, while the SADS for the clad produced at the same processing parameters but forced cooling in liquid nitrogen bath was 1.28 µm, indicating a significant difference in the cooling and solidification rates. The fine-grained microstructure is preferred because according to the relationship of Hall–Petch, the lower grain size increases the microhardness and also increases the wear resistance. Moreover, the optical micrographs show dispersive dots that are mainly in the interdendritic regions, which most likely are precipitations of complex carbides, as shown in Figure 14c,d.

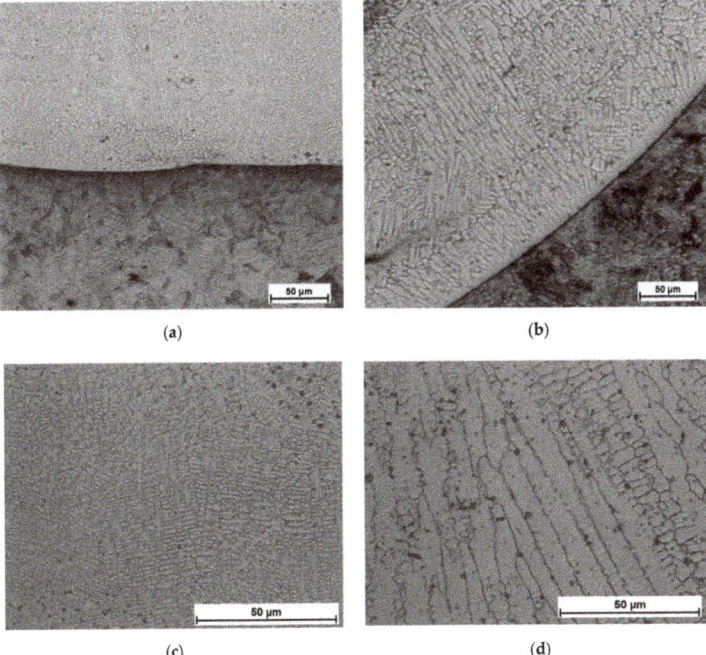

Figure 14. Microstructure of the comparative stringer beads LC4 and HC4 produced at the same process parameters (scanning speed 250 mm/min, laser output power 1000 W, energy input 240 J/mm) but different cooling conditions (Tables 3 and 5): (**a**,**c**) forced cooling under cryogenic conditions in a liquid nitrogen bath—interface region and middle region of the clad, respectively; and (**b**,**d**) free cooling in ambient air—interface region and middle region of the clad, respectively.

To more accurately determine the microstructure, SEM micrographs at high magnification were analyzed. The following is an example of the microstructure for the clad layer HC4 produced under cryogenic conditions. As can be seen in Figure 15, the typical very fine martensitic microstructure with clear retained austenite was disclosed within the dendritic areas, while highly dispersed nanometric-sized (200–360 nm) precipitations can be found in the interdendritic regions. Further EDS analysis indicated that the nanometric-sized precipitations are composed of tungsten, molybdenum vanadium, chromium, and iron, as shown in Figure 16. These results of SEM and EDS analysis clearly indicate that the dendritic grains are composed of solid solution based on Fe (Figure 16a,c), while the interdendritic regions are most likely composed of eutectic and precipitations of fine complex carbides (Figure 16a,b), which were identified by XRD analysis as $M_{23}C_6$ and M_7C_3, as shown in Figure 12.

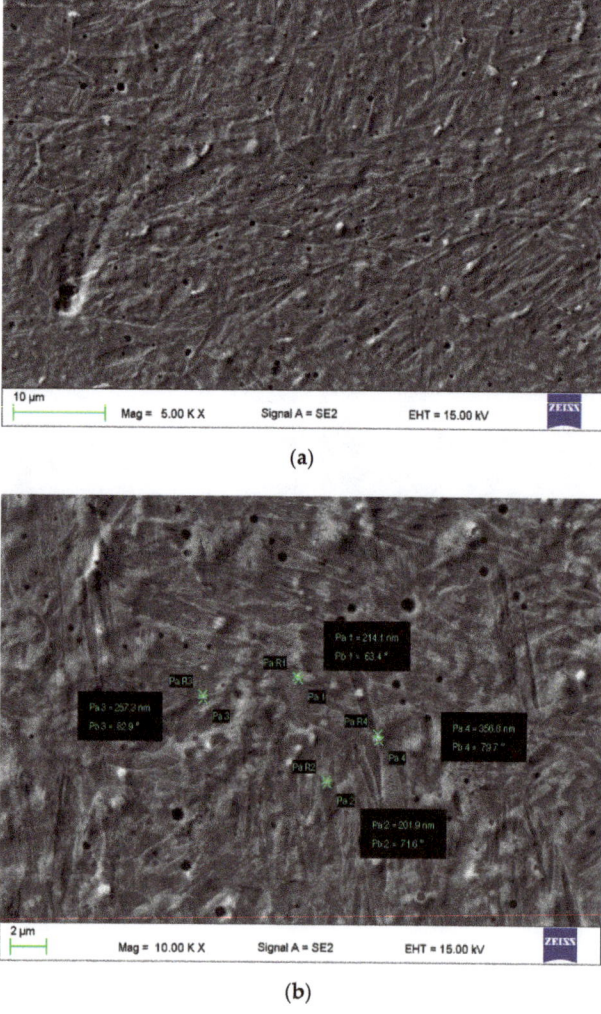

Figure 15. SEM micrograph in the middle region of the clad of stringer bead HC4 produced by the laser deposition of Fe-based powder (scanning speed 250 mm/min, laser output power 1000 W, energy input 240 J/mm) at forced cooling under cryogenic conditions in a liquid nitrogen bath (Table 3, Figures 12 and 16): (**a**) an overview of the microstructure within the dendritic regions (typical martensitic morphology); and (**b**) a close view of precipitations in interdendritic regions with measuring the dimensions (201.9–356.8 nm).

Table 5. Results of determination of the secondary dendrite arm spacing (SDAS) for the comparative stringer beads LC4 and HC4 produced at the same process parameters (scanning speed 250 mm/min, laser output power 1000 W, energy input 240 J/mm) but different cooling conditions (Table 3, Figure 14c,d).

Surface Layer	Measurement Number	Length (μm)	Total Number of Dendrite Arms	Individual Value of SADS (μm)	Mean Value of SADS (μm)	Standard Deviation of SADS (μm)
LC4	1	50	18	2.78	3.10	0.38021
	2	70	23	3.04		
	3	70	19	3.68		
	4	70	21	3.33		
	5	50	19	2.63		
	6	50	16	3.13		
HC4	1	37.5	28	1.34	1.28	0.09554
	2	22.5	20	1.13		
	3	25	19	1.32		
	4	30	22	1.36		
	5	22.5	17	1.32		
	6	25	21	1.19		

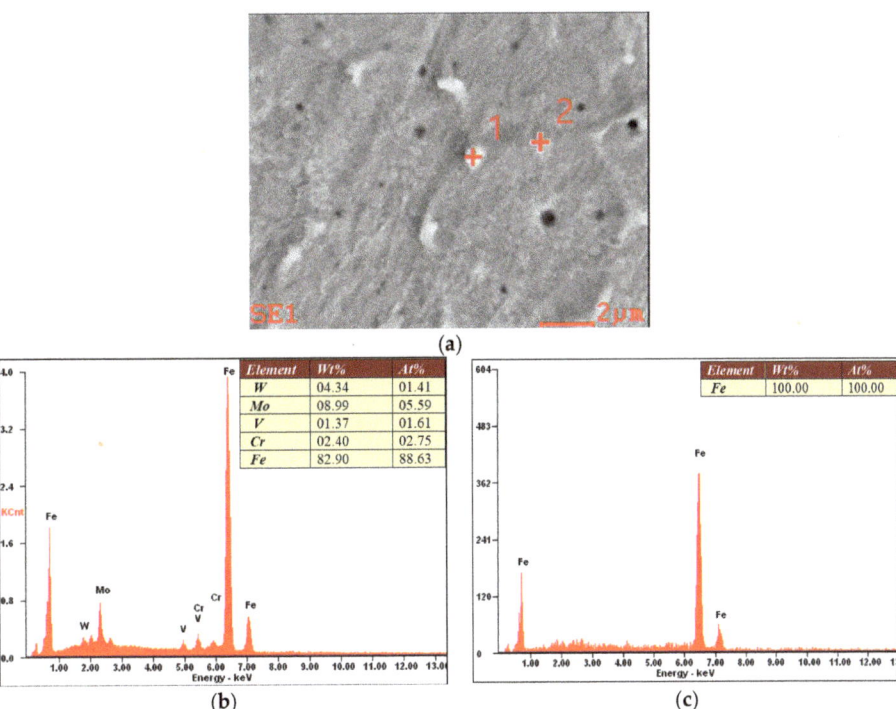

Figure 16. SEM micrograph (**a**) of the middle region of the clad of stringer bead HC4 (Table 3) chosen for energy-dispersive spectrometer (EDS) analysis: (**b**) EDS spectrum of the precipitation (a white dot in point No. 1); (**c**) EDS spectrum of the dendrite area (a gray region below point No. 2).

4. Conclusions

The novel technique of laser deposition of Fe-based powder under cryogenic conditions provided by the liquid nitrogen bath was successfully demonstrated. Due to intensive evaporation of the liquid nitrogen and the presence of nitrogen vapors in the melt pool region, the proposed technique can be considered as a hybrid process combining laser powder deposition (conventional laser cladding) with laser alloying by the gaseous nitrogen (precisely, laser gas nitriding—LGN).

The results of the study indicate that due to the supercooled substrate, which dissipates the absorbed laser energy, the penetration depth, width of HAZ, as well as the dilution (49% for cladding at free cooling and 5.2% under cryogenic conditions for comparative samples LC4 and HC4) can be significantly reduced. Such low rates of dilution provide beneficial profiles of hardness, uniform distribution in the clad layer with highest values of hardness (740–800 HV1 for clad layers produced under cryogenic conditions, and 370–780 HV1 for clad layers produced at free cooling).

Moreover, the test stringer beads produced under cryogenic conditions tend to increase in height and decrease in width, which can especially beneficial in the case of single-layer coatings of different composition than the substrate. Therefore, the technique of forced cooling can be useful for shaping the geometry of the clad layers. It was also found that forced cooling under cryogenic conditions has a beneficial effect on the refinement of the microstructure of the clad layers produced by the Fe-based powder, which was quantitatively determined by measuring the secondary dendrite arm spacing (SADS 3.1 µm for LC4 and 1.28 µm for HC4).

The above features such as the high hardness, homogeneity, and refinement of the microstructure may have also a beneficial effect on the tribological properties of coatings produced under forced cooling. However, further comprehensive stud related to the influence of forced cooling on thermal cycles, cooling, and solidification rates, as well as the wear characteristic of coatings under different load, is necessary.

Author Contributions: Conceptualization, methodology, investigation, formal analysis, validation, supervision, writing—review and editing, A.L.; project administration, funding acquisition, D.Ś. All authors have read and agreed to the published version of the manuscript.

Funding: The research was supported by the National Centre for Research and Development, Poland, under the grant number POIR.01.01.01-00-0278/15-007-02, financed by EU funds.

Acknowledgments: The authors thank all those who assisted in the preparation of samples and the conducting of measurements for their technical support, especially to Marcin Żuk, Adrian Kukofka, and their students who assisted in the laboratory tests.

Conflicts of Interest: The authors declare no conflict of interest.

References

1. Kusinski, J.; Kąc, S.; Kopia, A.; Radziszewska, A.; Rozmus-Górnikowska, M.; Major, B.; Major, L.; Marczak, J.; Lisiecki, A. Laser modification of the materials surface layer—A review paper. *Bull. Pol. Acad. Sci. Tech. Sci.* **2012**, *60*, 711–728. [CrossRef]
2. Gopinath, M.; Thota, P.; Kumar, A. Role of molten pool thermo cycle in laser surface alloying of AISI 1020 steel with in-situ synthesized TiN. *Surf. Coat. Technol.* **2019**, *362*, 150–166. [CrossRef]
3. Lisiecki, A. Mechanisms of hardness increase for composite surface layers during laser gas nitriding of the Ti6A14V alloy. *Mater. Technol.* **2017**, *51*, 577–583.
4. Kik, T.; Górka, J. Numerical simulations of laser and hybrid S700MC T-joint welding. *Materials* **2019**, *12*, 516. [CrossRef] [PubMed]
5. Lisiecki, A. Welding of titanium alloy by different types of lasers. *Arch. Mater. Sci. Eng.* **2012**, *58*, 209–218.
6. Lisiecki, A.; Kurc-Lisiecka, A. Erosion wear resistance of titanium-matrix composite Ti/TiN produced by diode-laser gas nitriding. *Mater. Teh.* **2017**, *51*, 29–34. [CrossRef]
7. Zhao, X.; Zhang, P.; Wang, X.; Chen, Y.; Liu, H.; Chen, L.; Sheng, Y.; Li, W. In-situ formation of textured TiN coatings on biomedical titanium alloy by laser irradiation. *J. Mech. Behav. Biomed.* **2018**, *78*, 143–153. [CrossRef]
8. Kurc-Lisiecka, A.; Lisiecki, A. Laser welding of new grade of advanced high strength steel Domex 960. *Mater. Teh.* **2017**, *51*, 199–204. [CrossRef]
9. Lisiecki, A. Titanium matrix composite Ti/TiN produced by diode laser gas nitriding. *Metals* **2015**, *5*, 54–69. [CrossRef]
10. Lisiecki, A. Study of optical properties of surface layers produced by laser surface melting and laser surface nitriding of titanium alloy. *Materials* **2019**, *12*, 3112. [CrossRef]

11. Kaźmierczak-Bałata, A.; Mazur, J. Effect of carbon nanoparticle reinforcement on mechanical and thermal properties of silicon carbide ceramics. *Ceram. Int.* **2018**, *44*, 10273–10280. [CrossRef]
12. Lisiecki, A.; Piwnik, J. Tribological characteristic of titanium alloy surface layers produced by diode laser gas nitriding. *Arch. Met. Mater.* **2016**, *61*, 543–552. [CrossRef]
13. Yang, C.; Liu, J. Intermittent vacuum gas nitriding of TB8 titanium alloy. *Vacuum* **2019**, *163*, 52–58. [CrossRef]
14. Wolowiec-Korecka, E.; Michalski, J.; Kucharska, B. Kinetic aspects of low-pressure nitriding process. *Vacuum* **2018**, *155*, 292–299. [CrossRef]
15. Lisiecki, A.; Burdzik, R.; Siwiec, G.; Konieczny, Ł.; Warczek, J.; Folęga, P.; Oleksiak, B. Disk laser welding of car body zinc coated steel sheets. *Arch. Met. Mater.* **2015**, *60*, 2913–2922. [CrossRef]
16. Pakieła, W.; Tański, T.; Pawlyta, M.; Brytan, Z.; Sroka, M. The structure and mechanical properties of AlMg5Si2Mn alloy after surface alloying by the use of fiber laser. *Appl. Phys. A* **2018**, *124*, 263. [CrossRef]
17. Bonek, M. The investigation of microstructures and properties of high speed steel HS6-5-2-5 after laser alloying. *Arch. Metall. Mater.* **2014**, *59*, 1647–1651. [CrossRef]
18. Tomków, J.; Rogalski, G.; Fydrych, D.; Łabanowski, J. Advantages of the application of the temper bead welding technique during wet welding. *Materials* **2019**, *12*, 915. [CrossRef]
19. Moskal, G.; Grabowski, A.; Lisiecki, A. Laser remelting of silicide coatings on Mo and TZM alloy. *Solid State Phenom.* **2015**, *226*, 121–126. [CrossRef]
20. Janicki, D. Microstructure and sliding wear behaviour of in-situ TiC-reinforced composite surface layers fabricated on ductile cast iron by laser alloying. *Materials* **2018**, *11*, 75. [CrossRef]
21. Lisiecki, A. Comparison of titanium metal matrix composite surface layers produced during laser gas nitriding of Ti6Al4V alloy by different types of lasers. *Arch. Met. Mater.* **2016**, *61*, 1777–1784. [CrossRef]
22. Li, C.; Zhang, Q.; Wang, F.; Deng, P.; Lu, Q.; Zhang, Y.; Li, S.; Ma, P.; Li, W.; Wang, Y. Microstructure and wear behaviors of WC-Ni coatings fabricated by laser cladding under high frequency micro-vibration. *Appl. Surf. Sci.* **2019**, *485*, 513–519. [CrossRef]
23. Lisiecki, A.; Ślizak, D.; Kukofka, A. Robotic fiber laser cladding of steel substrate with iron-based metallic powder. *Mater. Perform. Charact.* **2019**, *8*, 1202–1213. [CrossRef]
24. Górka, J.; Czupryński, A.; Żuk, M.; Adamiak, M.; Kopyść, A. Properties and structure of deposited nanocrystalline coatings in relation to selected construction materials resistant to abrasive wear. *Materials* **2018**, *11*, 1184. [CrossRef]
25. Lukaszkowicz, K.; Jonda, E.; Sondor, J.; Balin, K.; Kubacki, J. Characteristics of the AlTiCrN+DLC coating deposited with a cathodic arc and the PACVD process. *Mater. Teh.* **2016**, *50*, 175–181. [CrossRef]
26. Klimpel, A.; Dobrzański, L.A.; Lisiecki, A.; Janicki, D. The study of properties of Ni-WC wires surfaced deposits. *J. Mater. Process. Technol.* **2005**, *164–165*, 1046–1055. [CrossRef]
27. Ji, X.; Qing, Q.; Ji, C.; Cheng, J.; Zhang, Y. Slurry erosion wear resistance and impact-induced phase transformation of titanium alloys. *Tribol. Lett.* **2018**, *66*, 64. [CrossRef]
28. Li, S.; Li, C.; Deng, P.; Zhang, Y.; Zhang, Q.; Sun, S.; Yan, H.; Ma, P.; Wang, Y. Microstructure and properties of laser-cladded bimodal composite coatings derived by composition design. *J. Alloys Compd.* **2018**, *745*, 483–489. [CrossRef]
29. Klimpel, A.; Dobrzański, L.A.; Lisiecki, A.; Janicki, D. The study of the technology of laser and plasma surfacing of engine valves face made of X40CrSiMo10-2 steel using cobalt-based powders. *J. Mater. Process. Technol.* **2006**, *175*, 251–256. [CrossRef]
30. Sadeghimeresht, E.; Markocsan, N.; Nylén, P. A comparative study of corrosion resistance for HVAF-sprayed Fe- and Co-based coatings. *Coatings* **2016**, *6*, 16. [CrossRef]
31. Wang, H.; Zhang, S.; Zhang, C.; Wu, C.; Zhang, J.; Abdullah, A. Effects of V and Cr on laser cladded Fe-based coatings. *Coatings* **2018**, *8*, 107. [CrossRef]
32. Zhu, H.; Li, Y.; Li, B.; Zhang, Z.; Qiu, C. Effects of low-temperature tempering on microstructure and properties of the laser-cladded AISI 420 martensitic stainless steel coating. *Coatings* **2018**, *8*, 451. [CrossRef]
33. Chen, J.; Zhou, Y.; Shi, C.; Mao, D. Microscopic analysis and electrochemical behavior of Fe-based coating produced by laser cladding. *Metals* **2017**, *7*, 435. [CrossRef]
34. Li, J.; Zhu, Z.; Peng, Y.; Shen, G. Effect of Carbon fiber addition on the microstructure and wear resistance of laser cladding composite coatings. *Coatings* **2019**, *9*, 684. [CrossRef]
35. Hou, X.; Du, D.; Chang, B.; Ma, N. Influence of scanning speed on microstructure and properties of laser cladded Fe-based amorphous coatings. *Materials* **2019**, *12*, 1279. [CrossRef]

36. Zhang, X.; Zhou, Y. Effect of deep cryogenic treatment on microstructure and wear resistance of LC3530 Fe-based laser cladding coating. *Materials* **2019**, *12*, 2400. [CrossRef] [PubMed]
37. Zieliński, A.; Jażdżewska, M.; Naroźniak-Łuksza, A.; Serbiński, W. Surface structure and properties of Ti6Al4V alloy laser melted at cryogenic conditions. *J. Achiev. Mater. Manuf. Eng.* **2006**, *18*, 423–426.
38. Zieliński, A.; Serbiński, W.; Majkowska, B.; Jażdżewska, M.; Skalski, I. Influence of laser remelting at cryogenic conditions on corrosion resistance of non-ferrous alloys. *Adv. Mater. Sci.* **2009**, *9*, 21–28. [CrossRef]
39. Majkowska, B.; Serbiński, W. Cavitation wearing of the SUPERSTON alloy after laser treatment at cryogenic conditions. *Solid State Phenom.* **2010**, *165*, 306–309. [CrossRef]
40. Cui, Z.; Shi, H.; Wang, W.; Xu, B. Laser surface melting AZ31B magnesium alloy with liquid nitrogen-assisted cooling. *Trans. Nonferrous Met. Soc. China* **2015**, *25*, 1446–1453. [CrossRef]

© 2020 by the authors. Licensee MDPI, Basel, Switzerland. This article is an open access article distributed under the terms and conditions of the Creative Commons Attribution (CC BY) license (http://creativecommons.org/licenses/by/4.0/).

Review

A Compact Review of Laser Welding Technologies for Amorphous Alloys

Jian Qiao [1], Peng Yu [1], Yanxiong Wu [2], Taixi Chen [3], Yixin Du [1] and Jingwei Yang [1,*]

1. School of Mechatronic Engineering and Automation, Foshan University, Foshan 528000, China; qiaoj@fosu.edu.cn (J.Q.); 2111851018@fosu.edu.cn (P.Y.); 2112052109@stu.fosu.edu.cn (Y.D.)
2. School of Physics and Optoelectronic Engineering, Foshan University, Foshan 528000, China; wuyanxiong@fosu.edu.cn
3. Ji Hua Laboratory, Foshan 528000, China; chentx@jihualab.com
* Correspondence: mejwyang@fosu.edu.cn; Tel.: +86-15989111040

Received: 24 November 2020; Accepted: 17 December 2020; Published: 18 December 2020

Abstract: Amorphous alloys have emerged as important materials for precision machinery, energy conversion, information processing, and aerospace components. This is due to their unique structure and excellent properties, including superior strength, high elasticity, and excellent corrosion resistance, which have attracted the attention of many researchers. However, the size of the amorphous alloy components remains limited, which affects industrial applications. Significant developments in connection with this technology are urgently needed. Laser welding represents an efficient welding method that uses a laser beam with high energy-density for heating. Laser welding has gradually become a research hotspot as a joining method for amorphous alloys due to its fast heating and cooling rates. In this compact review, the current status of research into amorphous-alloy laser welding technology is discussed, the influence of technological parameters and other welding conditions on welding quality is analyzed, and an outlook on future research and development is provided. This paper can serve as a useful reference for both fundamental research and engineering applications in this field.

Keywords: laser welding; amorphous alloy; weld quality

1. Introduction

Amorphous alloys are an emerging class of advanced materials. Compared with conventional metals, their internal atoms are in a (long-range) disordered state. In addition, amorphous alloys have excellent mechanical, physical, and chemical properties, such as high elasticity, high permeability, and high strength and hardness [1]. Some of them have been selected as ideal materials for many industrial and civil applications, for example structural material, sporting goods, wear-/corrosion-resistant coatings, and transformers. In the 1960s, Klement et al. [2] first synthesized and investigated an Au–Si amorphous alloy at the California Institute of Technology. This research was further advanced by Chen et al. [3] during the 1970s. Subsequently, the group of Chen succeeded in producing a millimeter-level Pd-based bulk amorphous alloy [4]. In the late 1980s, research groups led by professors Inoue (Japan) and Johnson (USA) opened up new research directions for the field of amorphous alloys. Both groups produced important breakthroughs for the development and preparation of bulk amorphous alloys (amorphous alloys above millimeter-level size). This led to substantial improvements in the amorphous forming ability of this alloy system and the successful preparation of alloys like La–Al–TM [5–7], Mg–La–TM [8], Zr–Al–TM [9], Pd–Cu–Ni–P [10], and Zr–Al–Cu–TM [11,12] (TM denotes a transition metal), as well as a series of bulk amorphous alloy systems. Since then, the field of amorphous alloys has experienced renewed interest, and a series of new amorphous alloy systems has been developed, which include Zr [13–15], Pd [16,17], Cu [18,19],

Fe [20,21], Ti [22,23], Mg [24,25], Ni [26,27], and Zn-based [28] systems. As the preparation techniques and composition designs continue to advance, new possible amorphous alloy systems continue to be found, which also increases their range of industrial applications.

While amorphous alloys have excellent properties such as high elasticity, high permeability, and high strength and hardness that make them suitable for many industrial applications, there are also some disadvantages. For example, their preparation requires a very high cooling rate, and there are tight limits with respect to forming dimensions. In addition, high heating rates are required in order to keep the amorphous structure, especially for welding. Solving the problem of the size limitation is crucial to increase the range of applications. Many researchers have studied and developed new methods to overcome these limitations, such as additive manufacturing [29,30], thermoplastic forming [31], and welding techniques. Laser welding, in particular, would be very suitable as it represents an efficient and precise welding method that uses a high energy-density laser beam for heating/melting. Other advantages of this technique are the fast welding speed, a stable welding process, high heating and cooling speeds, and a small heat-affected zone (HAZ). Compared to other welding methods, laser welding has important advantages for amorphous alloys.

This review aims to outline the latest research and existing problems for the laser welding of metallic glass. The important conclusions about laser welding of amorphous alloys are summarized at the end of this paper.

2. Laser Welding of Amorphous Alloys

2.1. Introduction to Laser Welding

Laser welding uses a high-energy laser beam to irradiate the surface of the materials. The temperature rises rapidly and evaporation leads to the formation of small holes. Heat transfers from the outer wall of the high-temperature cavity, which causes the metal surrounding the cavity to melt. As the laser beam moves, molten metal flows through the keyhole and rejoins at the original interface to form a weld, and the two materials connect. Figure 1 shows a schematic illustrating the principle of laser welding.

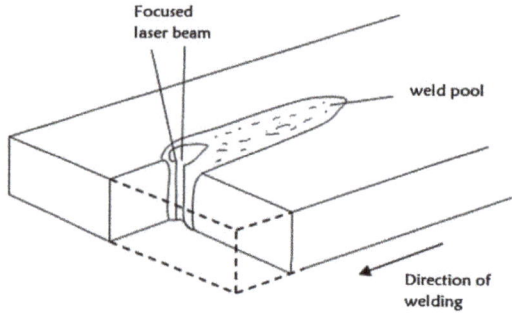

Figure 1. Schematic to illustrate laser welding.

2.2. Effect of Process Parameters on the Weld Quality

Thermodynamically, amorphous alloys are metastable and prone to changes in the microstructure by external factors. Once crystallization occurs during the melting and solidification of an amorphous alloy, the original excellent properties will be lost and lead to poor welding quality. Therefore, the evaluation of weld quality considers not only the geometry and performance of the weld but also whether the material maintains the amorphous structure. Kim et al. [32,33] used a Nd:YAG pulsed laser to weld 1-mm-thick $Cu_{54}Ni_6Zr_{22}Ti_{18}$ bulk amorphous alloy. The microstructure of the weld was investigated using scanning electron microscopy (SEM), micro-area X-ray diffractometry (XRD), and transmission electron microscopy (TEM). First, they analyzed the impact of the pulse energy on the welding quality and found that the larger the pulse energy, the easier it was to crystallize in the HAZ. Subsequently, the effect of frequency and speed on the welding quality was studied. Their results revealed that lower speeds and higher frequencies lead to crystallization within the HAZ. Finally, a weld without crystallization was obtained for a peak power of 1000 W, a pulse width of 4 ms, a frequency of 2 Hz and a speed of 60 mm/min. Louzguine-Luzgin et al. [34] used a fiber laser to weld a $Ni_{53}Nb_{20}Ti_{10}Zr_8Co_6Cu_3$ amorphous ribbon with a thickness of 25 micrometers. The two samples were successfully connected at a speed of 6 m/min, 120 W power, and a wavelength of λ = 1.07 µm. As shown in Figure 2c, the high-resolution TEM image of the sample's weld area consists of a wider halo and a diffraction ring, which reveals a typical amorphous structure. However, some weak spot-like features within the amorphous halo suggest some medium-range order, which typically lies within the 0.5–1 nm length scale. In the SEM micrographs of certain areas of the polished part of the weld (Figure 2b), some particles were found to be embedded in the glassy matrix. These ZrO_2 particles were confirmed using selected area electron diffraction (SAED) patterns—see Figure 2a. The amorphous structure of the welding sample did not change. However, to find out whether the ZrO_2 particles affect the quality of the weld requires further study. This study also shows that it is necessary to provide good protection from oxidation during laser welding of amorphous alloys.

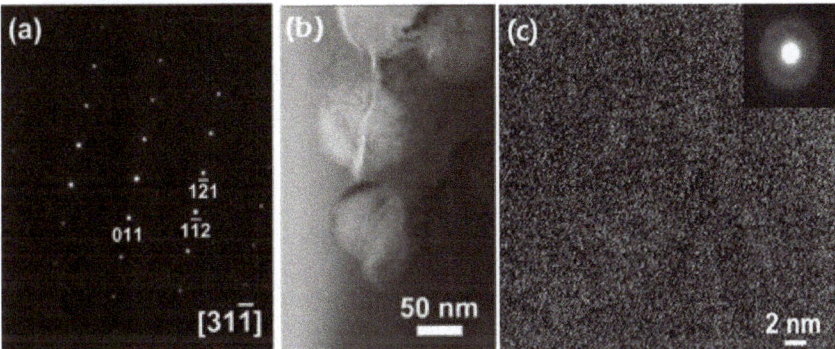

Figure 2. (a) Selected area electron diffraction (SAED) pattern for a particle in bright-field transmission electron microscopy (TEM) imaging mode. (b) Scanning electron microscopy (SEM) micrograph of the same sample. (c) High-resolution TEM image of the weld area [34] (with the permission of Elsevier).

Kawahito et al. [35] studied the feasibility of laser welding of Zr-based amorphous alloys. The group used a fiber laser with a laser power of 2.5 kW to weld a 1-mm-thick $Zr_{55}Al_{10}Ni_5Cu_{30}$ bulk amorphous alloy. The effect of different speeds on the crystallization of welds was studied. As a result, the fiber laser beam could be used to weld Zr-based metallic glass at an ultra-high speed at 72 m/min and still maintain the amorphous state. Huei-Sen Wang et al. used a Nd:YAG laser to weld $(Zr_{53}Cu_{30}Ni_9Al_8)Si_{0.5}$ [36] and $(Zr_{44}Cu_{36}Al_8Ag_8Ta_4)Si_{0.75}$ [37] bulk metallic glasses (BMGs). The microstructure evolution during laser-welding was studied by changing peak power, pulse width, and pulse energy. As shown in Figure 3, crystallization was observed in the HAZ at high pulse-energy but no crystallization occurred in the weld fusion zone (WFZ). Figure 4a shows a higher magnification of the crystalline phase in the HAZ, which further reveals the crystalline phase with a particle size of 30–200 nm in the precipitate (see Figure 4b). The particle size of the crystalline phase around the precipitated phase was about 20–30 nm (see Figure 4c). In addition, it can be observed (from the Figure 4a) that cracks formed in the central region of the precipitates, where the grain size of the crystalline phase exceeded 50 nanometers. The microstructure of the crystals in the HAZ is described in detail, which improves understanding of the crystallization properties of laser welding amorphous alloys. Pilarczyk et al. [38] studied the weld-surface morphology and nano-mechanical properties of

Zr-based laser welded samples. As shown in Figure 5, a detailed analysis of the surface morphology indicates that the molten zone has a smooth surface, while the HAZ shows a slight roughness for the crystalline phase. Furthermore, a hardness test showed the nano-hardness of the HAZ below that of the smooth molten zone.

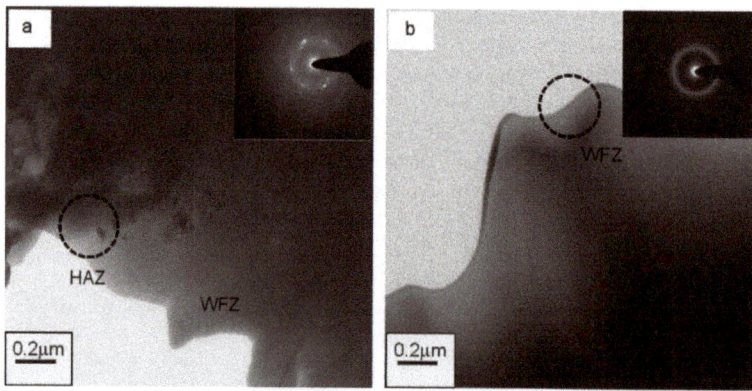

Figure 3. TEM bright-field images and selected area diffraction patterns for: (**a**) heat-affected zone (HAZ), (**b**) weld fusion zone (WFZ) [36] (with the permission of Elsevier).

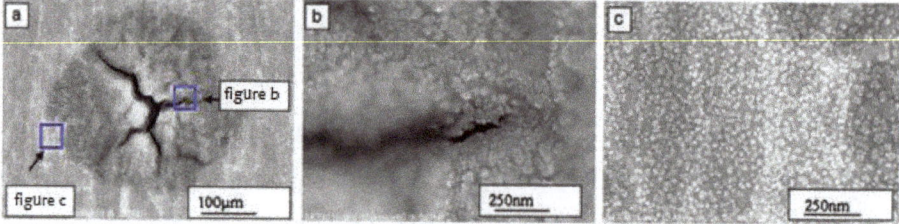

Figure 4. (**a**) Higher magnification of the crystalline phase within the HAZ, (**b**) crystalline phases in the central area of the precipitates with a particle size of 30–200 nm, (**c**) crystalline particles with a particle size of 20–30 nm [36] (with the permission of Elsevier).

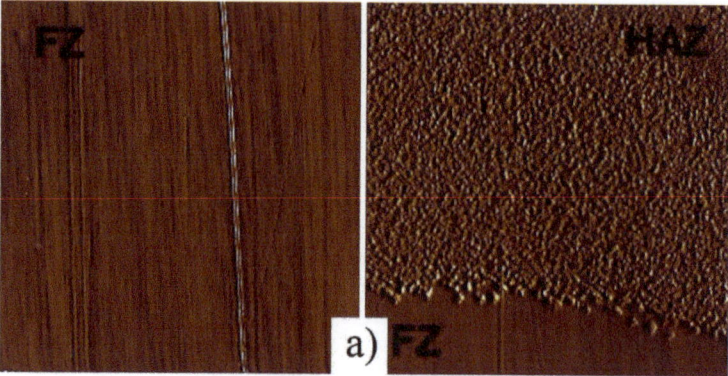

Figure 5. *Cont.*

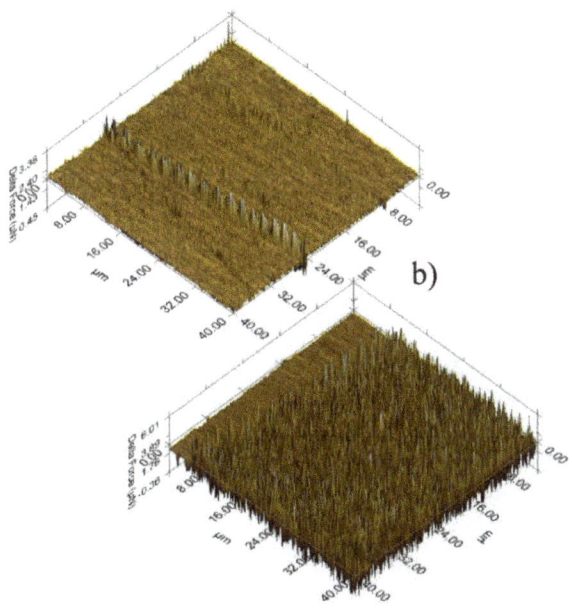

Figure 5. The surface morphology of a Zr–Cu–Ni–Al laser weld in the fusion zone (FZ) and the HAZ: (a) 2D image, (b) 3D image [38] (with the permission of Polish Academy of Sciences).

G. Wang et al. [39] successfully welded an amorphous alloy $Ti_{40}Zr_{25}Ni_3Cu_{12}Be_{20}$ block with a thickness of 3 mm using laser welding. The laser power was 3.5 kW, the spot size was 0.2 mm, and the welding speed was 6 m/min to 10 m/min. It was concluded that, as the speed decreased, the degree of crystallization in the weld increased, and the tensile strength of the joint decreased. Minghua Chen et al. [40] used a pulsed Nd:YAG laser to weld β-Ti-dendrite-enhanced amorphous alloy. The effects of laser excitation current, pulse width, and welding speed on the microstructure of the joint were studied. The microstructures of the welding joints were observed and analyzed. The microstructure in the fusion zone (FZ) consisted of β-Ti grains and an amorphous matrix. As shown in Figure 6a, it was found that the size of the Ti particles in the melting zone was much smaller than in the base material. This caused the hardness of the melting zone to be slightly higher than that in the base material. This, again, is due to the rapid cooling rate and the impact of laser pulses on the molten zone. It was concluded that the increase of the laser-pulse excitation current introduced a strong shock-wave into the molten pool, which led to finer grains in the molten zone—see Figure 6b,c. The growth state of the grains in the weld was studied in detail, which can provide experimental evidence and guidance to improve laser-welding of amorphous alloys.

Some researchers have studied the laser welding process of Pd-base [41] and Fe-base [42,43] amorphous alloys. Similar to previous research, the amorphous properties of the HAZ and melting zone of the welded joint were determined using differential scanning calorimetry (DSC) and XRD. The microstructure and surface morphology of the weld joint was examined using X-ray analysis, atomic force microscopy, and high-resolution transmission electron microscopy. As shown in Figure 7, the X-ray diffraction pattern of the FZ showed wide fuzzy spectra, which are characteristic for amorphous structures. In addition, the surface morphology of the HAZ revealed a slightly rough characteristic, and the X-ray diffraction pattern showed small diffraction lines from the crystal phase, which indicate that the area was amorphous–crystalline. Furthermore, the nano-hardness of the melting zone was the same as that of the base material, while the nano-hardness of the HAZ was a little lower. This is consistent with previous research results of Zr-based amorphous alloys [38].

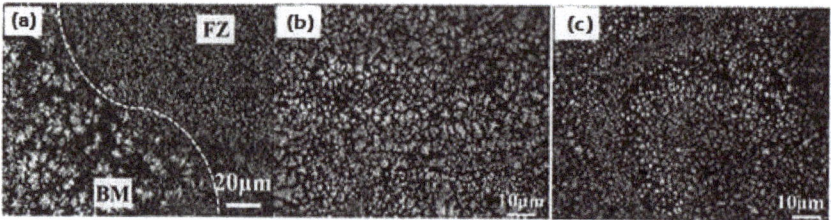

Figure 6. (a) Microstructure of the joints. Grains in the transition region in the weld joint: (**b**) laser-pulse electric current of 80 A and laser pulse energy of 6.0 J; (**c**) laser-pulse electric current of 100 A and laser pulse energy of 8.6 J [40] (with the permission of Elsevier).

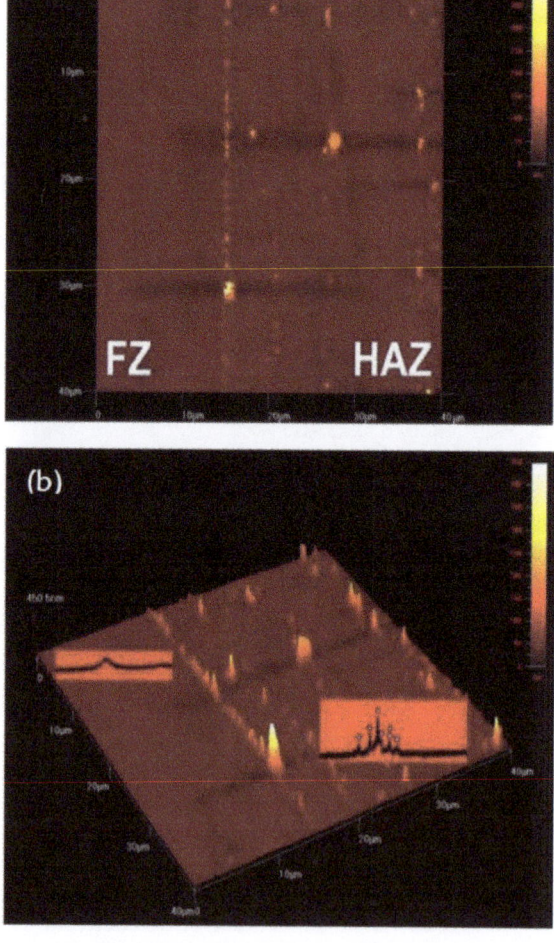

Figure 7. Surface morphology of the FZ and HAZ: (**a**) 2D image, (**b**) 3D image made by atomic force microscopy (AFM) with X-ray diffraction pattern [43] (with the permission of Polish Academy of Sciences).

In recent years, researchers have used different laser systems to conduct welding process research for different metal-glass systems. It can be found from various studies that whether the amorphous alloy structure remains amorphous is the most important condition to measure the welding quality. Lower welding speed, higher pulse energy or laser power, and larger pulse width will lead to crystallization of amorphous alloy welding samples, thus affecting the welding quality.

2.3. Effect of Other Welding Conditions on the Weld Quality

Laser welding of amorphous alloys is a very complex process. There are several other welding parameters that affect the crystallization of the amorphous alloy in addition to the process parameters. For example, Huei-Sen Wang et al. [44,45] studied the effect of the initial welding temperature on the weld quality, which was altered using a liquid cooling device. The microstructure evolution and mechanical properties of the welded samples were analyzed using a combination of SEM, TEM, DSC, and a Vickers microhardness test. It was found that, at low laser-energy and initial welding temperature (at 0 °C), no crystallization occurred within the HAZ. In addition, the microhardness was similar to that of the base material. When the initial welding temperature was increased, a crystallized precipitate appeared in the heat-affected zone, and cracks were found in the crystallized precipitate in the crystallized zone. These cracks decreased the microhardness. When the initial welding temperature was 0 °C, no crystallization occurred in either the HAZ or the melting zone, and the properties of the welding sample did not change compared to the base material. Later research [46] confirmed this conclusion. When the initial welding temperature was 0 °C, no crystallization occurred in the HAZ or the melting zone, and the properties of the welding sample did not change compared to the base material. This confirmed that decreasing the initial welding temperature can improve the weldability of amorphous alloy.

Chen Biao et al. [47] used a laser to weld annealed $Zr_{55}Cu_{30}Al_{10}Ni_5$ bulk amorphous alloy. The group studied the effect of the annealing process and welding parameters on the weld quality. As shown in Figure 8, the unannealed weld sample was completely crystallized, while the annealed weld sample retained some amorphous features. When the annealing temperature was 415 °C, and the annealing time was 10 min, the SAED pattern of the weld sample showed a typical amorphous diffraction ring together with several bright specks—see Figure 8a. According to the above description, the sample maintained its amorphous structure and only a few nanocrystals were present in the sample substrate. Interestingly, due to the presence of nanocrystals, the welded joints of the sample had a higher microhardness and flexural strength than the unannealed samples. Therefore, it can be concluded that a suitable annealing temperature and time can improve the laser welding quality of Zr-based amorphous alloys.

According to a study by Pilarczyk [48], the metallurgical properties of amorphous alloys before welding have a great impact on the welding quality. Later, Huei-Sen Wang et al. [49] and Yiyu Shen et al. [50] conducted further research of this phenomenon. Because of the continuity of temperature during amorphous alloy casting, pre-existing nuclei cannot be avoided in any amorphous alloy casting process. Therefore, Yiyu Shen et al. investigated the effect of pre-existing nuclei on the crystallization of laser-welded Zr-based amorphous alloys. According to classical nucleation/growth theory, the evolution of a crystalline phase during laser welding of amorphous alloys was analyzed. The results show that the samples with higher pre-existing nuclear density were more likely to crystallize under the same laser welding conditions. The nuclear density of amorphous-alloy castings can be reduced by increasing the cooling rate during casting. Therefore, increasing the cooling rate improves laser welding of amorphous alloy casting parts.

It is useful to change other laser welding conditions to avoid crystallization in the welding process. Reducing the initial welding temperature will increase the cooling rate and reduce the amount of crystal precipitation in the heat-affected zone, which is beneficial to improve the welding quality. In addition, appropriate annealing temperature and annealing time are beneficial to improve the weld quality, micro-hardness, and bending strength of the welded area.

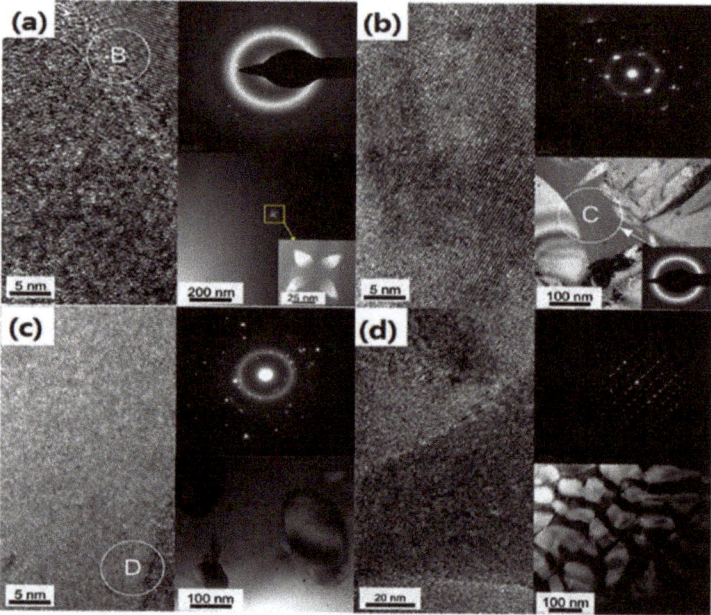

Figure 8. TEM images of the welding joints for different annealing conditions (**a**) annealing temperature: 415 °C, annealing time: 10 min (**b**) annealing temperature: 415 °C, annealing time: 30 min (**c**) annealing temperature: 430 °C, annealing time: 10 min, (**d**) not annealed [47] (with the permission of Elsevier).

2.4. Laser Welding of Amorphous Alloys and Crystalline Metals

To expand the industrial application range of amorphous alloys, many researchers began to search for a welding method to combine amorphous alloys with conventional crystal metals. Chen Biao et al. [51] used an optical fiber laser to weld Zr-based bulk metallic glass and zirconium. The structure and quality of the welded joint were measured using SEM, TEM, and a hardness tester. Laser welding did not change the amorphous structure of BMG near the melting zone. At 1.3 kW and 8 m/min, no cracks or defects were observed at the welding joint, and the WFZ maintained the same hardness as the amorphous base material. The research revealed that laser welding can be applied to connect amorphous alloys with conventional metals. Alavi et al. [52] used the laser welding method to fill the gap of a 3-mm-thick plain carbon steel plate with Fe-based amorphous alloy powder. The 3-mm-thick steel substrate was successfully welded together, but the amorphous structure was not maintained in the laser bonding region. Pingjun Tao et al. [53] used a pulsed laser to weld $Zr_{55}Cu_{30}Ni_5Al_{10}$ plate-like BMG and alloy powder of the same composition. The surface layers of the BMG can be significantly intensified after the laser welding treatments. However, compared with the BMG matrix, the mechanical properties of HAZ are slightly lower.

Xiao Wang et al. [54] successfully welded Fe-based amorphous ribbon and copper foil via laser shock welding. The thickness of the Fe-based amorphous ribbon was 28 μm, and the thickness of the copper foil was 30 μm. A series of process experiments was conducted with different pulse energies. SEM micrographs of welding interfaces with different pulse energies are shown in Figure 9. The welding interface was straight for low pulse energy, while the welding interface was wavy at high pulse energy. The interface was subjected to greater impact and produced high plastic deformation and shear stress because of the high pulse energy. The amorphous structure of the Fe-based amorphous ribbon was not affected by laser welding when the pulse energy was 835 mJ. In addition, the joints exhibited a higher degree of hardness than before the welding. These results show that the Fe-based amorphous ribbon and crystalline copper were successfully welded using laser impact welding and a good welding quality could be maintained.

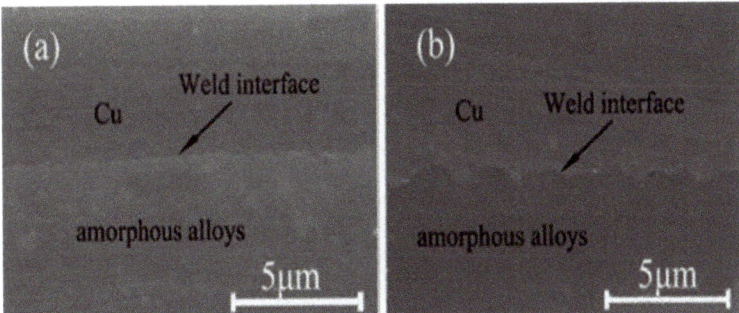

Figure 9. SEM micrographs of the welding interfaces obtained at different pulse energies: (a) 835 mJ, (b) 1550 mJ [54].

Xiaoquan Yu et al. studied the laser-welding process of Zr-based bulk amorphous alloy with a 1100 aluminum alloy plate [55] as well as a copper plate [56]. The microstructure and chemical composition of the joint were analyzed using SEM and XRD, and the results are shown in Figure 10. The results show that the amorphous base-material maintained the amorphous structure, while Zr_2Ni and Zr_2Al phases appeared in the melting zone, the heat affected zone, and the junction between the melting zone and the aluminum alloy plate. The hardness test showed that the Vickers hardness of the joint exceeded that of the amorphous base-material, and the maximum hardness occurred at the junction of the aluminum plate with the welded joint. In the welding samples of Zr-based metallic glass and copper plate, both the molten zone and the amorphous base material zone maintained an amorphous characteristic. However, a crystalline phase of Zr and Cu appeared at the junction of the molten zone with the copper plate. The hardness of the amorphous structure in the molten zone was improved compared to the amorphous base material. The research results above confirm the feasibility of laser welding of amorphous alloys with crystalline metals. The hardness of the welded sample joint was improved compared to the parent material, which represents a new approach to extend the range of industrial applications for amorphous alloys.

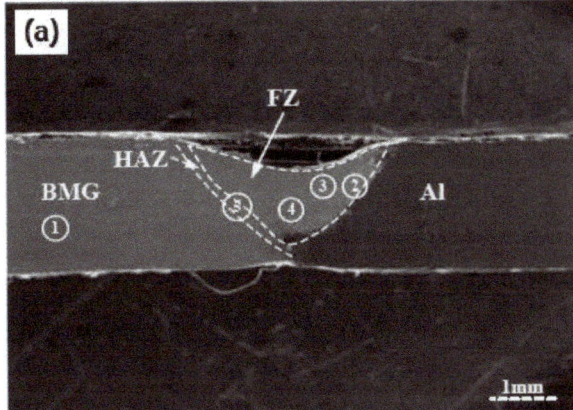

Figure 10. *Cont.*

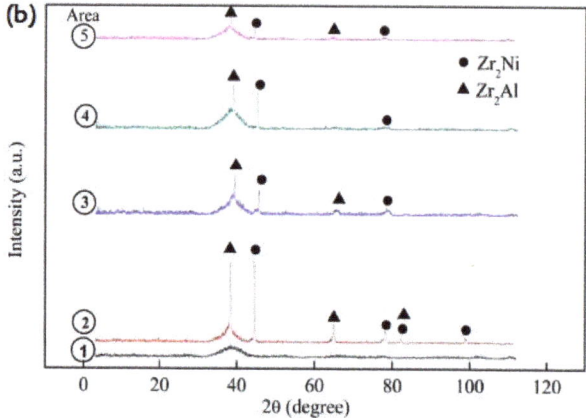

Figure 10. (a) SEM image of the joint, (b) micro-area X-ray diffractometry (XRD) patterns for the selected zones in the joint [55] (with the permission of IOP Publishing).

3. Conclusions

This review focused on the successful applications of laser welding technology for amorphous alloys. Laser welding, thanks to its high heating and cooling rates, is suitable for the welding of amorphous alloys. The technology is also very useful in solving the problem of size limitation for amorphous alloys in industrial applications, which expands the range of industrial applications significantly. Over the past decade, researchers have studied different laser systems to weld different amorphous alloys, the list is shown in Table A1 (Appendix A), and they have investigated the impact of process parameters and other welding conditions on the microstructure of amorphous alloys. They have also used several methods to characterize the quality of the welds. There have also been reports of laser welding between amorphous alloys and crystalline metals. Despite many reported studies, which provide valuable experimental support for the development and optimization of laser welding of metallic glass, much is still unknown.

The development of new laser-processing systems and amorphous alloy systems will continue to advance laser-welding technology, and many current obstacles could be overcome in the future. Because our understanding of the interaction between lasers and amorphous alloys is incomplete, it is necessary to study the involved mechanisms in more detail in the future. In addition, numerical simulations will be very useful, which can be used to assist in the design and optimization of laser welding of amorphous alloys. For example, by comparing the continuous heating transformation curve with the simulated temperature curve of the amorphous alloy, it is possible to predict when these interact, thus leading to crystallization during the welding process. To date, only a few studies have been done that used numerical simulations of laser welding of amorphous alloys. While this is a promising research direction, an accurate simulation is not a small challenge. Factors that need to be considered include vaporization, plasma effects, and any chemical composition changes in the materials. Finally, specific applications of laser-welded amorphous-alloy parts in industry should be investigated, which will promote the wider utilization of these materials in many industrial sectors.

Author Contributions: Conceptualization, J.Q. and Y.W.; writing—original draft preparation, P.Y. and Y.D.; writing—review and editing, J.Y. and T.C. All authors have read and agreed to the published version of the manuscript.

Funding: This research was funded by The National Natural Science Foundation of China (Grant No. 51805084), and the project funds support provided by Ji Hua Laboratory (Project No. X180081UZ180).

Conflicts of Interest: The authors declare no conflict of interest.

Appendix A

Table A1. Materials and laser system used of the currently reviewed articles.

Paper	Material	Laser System
Kim et al. [32,33]	$Cu_{54}Ni_6Zr_{22}Ti_{18}$	Nd: YAG laser (pulsed)
Louzguine-Luzin et al. [34]	$Ni_{53}Nb_{20}Ti_{10}Zr_8Co_6Cu_3$	fiber laser (cw)
Kawahito et al. [35]	$Zr_{55}Al_{10}Ni_5Cu_{30}$	fiber laser (cw)
Huei-Sen Wang et al. [36,37]	$(Zr_{53}Cu_{30}Ni_9Al_8)Si_{0.5}/(Zr_{44}Cu_{36}Al_8Ag_8Ta_4)Si_{0.75}$	Nd: YAG laser (pulsed)
Pilarczyk et al. [38]	$Zr_{55}Cu_{30}Ni_5Al_{10}$	TruLaser Station 5004
G. Wang et al. [39]	$Ti_{40}Zr_{25}Ni_3Cu_{12}Be_{20}$	Trumpf Trudisk 6002
Minghua Chen et al. [40]	TiZrBeCuMo	Nd: YAG laser (pulsed)
Ling Shao et al. [41]	$Pd_{43}Cu_{27}Ni_{10}P_{20}$	IPG Photonics 1500
Pilarczyk et al. [42,43]	$Fe_{37.44}Co_{34.56}B_{19.2}Si_{4.8}Nb_4$	TruLaser Station 5004
Huei-Sen Wang et al. [44,45]	$(Zr_{53}Cu_{30}Ni_9Al_8)Si_{0.5}$	Nd: YAG laser (pulsed)
Huei-Sen Wang et al. [46]	$(Zr_{53}Al_{17}Co_{29})_{99}Ta_1$	Nd: YAG laser (pulsed)
Chen Biao et al. [47]	$Zr_{55}Cu_{30}Al_{10}Ni_5$	fiber laser (cw)
Pilarczyk [48]	Fe-Co-B-Si-Nb	TruLaser Station 5004
Huei-Sen Wang et al. [49]	$(Zr_{48}Cu_{36}Al_8Ag_8)Si_{0.75}$	Nd: YAG laser (pulsed)
Yiyu Shen et al. [50]	$Zr_{52.5}Ti_5Al_{10}Ni_{14}Cu_{17.9}$	fiber laser (cw)
Chen Biao et al. [51]	$Zr_{41}Ti_{14}Cu_{12}Ni_{10}Be_{23}$ and zirconium metal	fiber laser (cw)
Alavi et al. [52]	carbon steel and Fe-based amorphous alloy powder	fiber laser (cw)
Pingjun Tao et al. [53]	$Zr_{55}Cu_{30}Ni_5Al_{10}$ and alloy powder of the same composition	pulsed laser
Xiao Wang et al. [54]	Fe-based amorphous alloys (GB1K101) and copper	Nd: YAG laser (pulsed)
Xiaoquan Yu et al. [55]	$Zr_{44}Ti_{11}Ni_{10}Cu_{10}Be_{25}$ and aluminum	Nd: YAG laser (pulsed)
Xiaoquan Yu et al. [56]	$Zr_{44}Ti_{11}Ni_{10}Cu_{10}Be_{25}$ and copper	Nd: YAG laser (pulsed)

References

1. Schroers, J. Processing of bulk metallic glass. *Adv. Mater.* **2010**, *22*, 1566–1597. [CrossRef] [PubMed]
2. Klement, W.; Willens, R.; Duwez, P. Non-crystalline structure in solidified gold–silicon alloys. *Nature* **1960**, *187*, 869–870. [CrossRef]
3. Chen, H.S.; Miller, C.E. A rapid quenching technique for the preparation of thin uniform films of amorphous solids. *Rev. Sci. Instrum.* **1970**, *41*, 1237–1238. [CrossRef]
4. Chen, H.S.; Krause, J.T.; Coleman, E. Elastic constants, hardness and their implications to flow properties of metallic glasses. *J. Non-Cryst. Solids* **1975**, *18*, 157–171. [CrossRef]
5. Inoue, A.; Kita, K.; Zhang, T.; Masumoto, T. An amorphous $La_{55}Al_{25}Ni_{20}$ alloy prepared by water quenching. *Mater. Trans. JIM* **1989**, *30*, 722–725. [CrossRef]
6. Inoue, A.; Zhang, T.; Masumoto, T. Al-La-Ni amorphous alloys with a wide supercooled liquid region. *Mater. Trans. JIM* **1989**, *30*, 965–972. [CrossRef]
7. Inoue, A.; Nakamura, T.; Sugita, T.; Zhang, T.; Masumoto, T. Bulky La–Al–TM (TM = Transition Metal) amorphous alloys with high tensile strength produced by a high-pressure die casting method. *Mater. Trans. JIM* **1993**, *34*, 351–358. [CrossRef]
8. Inoue, A.; Kohinata, M.; Tsai, A.P.; Masumoto, T. Mg–Ni–La amorphous alloys with a wide supercooled liquid region. *Mater. Trans. JIM* **1989**, *30*, 378–381. [CrossRef]
9. Inoue, A.; Zhang, T.; Masumoto, T. Zr-Al-Ni amorphous alloys with high glass transition temperature and significant supercooled liquid region. *Mater. Trans. JIM* **1990**, *31*, 177–183. [CrossRef]
10. Inoue, A.; Nishiyama, N.; Matsuda, T. Preparation of bulk glassy $Pd_{40}Ni_{10}Cu_{30}P_{20}$ Alloy of 40 mm in diameter by water quenching. *Mater. Trans. JIM* **1996**, *37*, 181–184. [CrossRef]
11. Zhang, T.; Inoue, A. Density, thermal stability and mechanical properties of Zr–Ti–Al–Cu–Ni bulk amorphous alloys with high Al plus Ti concentrations. *Mater. Trans. JIM* **1998**, *39*, 857–862. [CrossRef]
12. Inoue, A.; Zhang, T. Fabrication of bulk glassy $Zr_{55}Al_{10}Ni_5Cu_{30}$ alloy of 30 mm in diameter by a suction casting method. *Mater. Trans. JIM* **1996**, *37*, 185–187. [CrossRef]
13. Peker, A.; Johnson, W.L. A highly processable metallic glass: $Zr_{41.2}Ti_{13.8}Cu_{12.5}Ni_{10.0}Be_{22.5}$. *Appl. Phys. Lett.* **1993**, *63*, 2342–2344. [CrossRef]
14. Lin, X.H.; Johnson, W.L. Formation of Ti-Zr-Cu-Ni bulk metallic glasses. *J. Appl. Phys.* **1995**, *78*, 6514–6519. [CrossRef]
15. Hays, C.C.; Schroers, J.; Geyer, U.; Bossuyt, S.; Stein, N.; Johnson, W.L. Glass forming ability in the Zr-Nb-Ni-Cu-Al bulk metallic glasses. *J. Metastable Nanocryst. Mater.* **2000**, *8*, 103–108. [CrossRef]

16. He, Y.; Schwarz, R.B.; Archuleta, J.I. Bulk glass formation in the Pd-Ni-P system. *Appl. Phys. Lett.* **1996**, *69*, 1861–1863. [CrossRef]
17. Nishiyama, N.; Inoue, A. Glass-forming ability of bulk $Pd_{40}Ni_{10}Cu_{30}P_{20}$ alloy. *Mater. Trans. JIM* **1996**, *37*, 1531–1539. [CrossRef]
18. Inoue, A.; Zhang, W.; Zhang, T.; Kurosaka, K. High-strength Cu based bulk glassy alloys in Cu-Zr-Ti and Cu-Hf-Ti ternary systems. *Acta Mater.* **2001**, *49*, 2645–2652. [CrossRef]
19. Dai, C.L.; Guo, H.; Shen, Y.; Li, Y.; Ma, E.; Xu, J. A new centimeter-diameter Cu-based bulk metallic glass. *Scr. Mater.* **2006**, *54*, 1403–1408. [CrossRef]
20. Ponnambalam, V.; Poon, S.J.; Shiflet, G.J. Fe-based bulk metallic glasses with diameter thickness larger than one centimeter. *J. Mater. Res.* **2004**, *19*, 1320–1323. [CrossRef]
21. Ponnambalam, V.; Poon, S.J.; Shiflet, G.J.; Keppens, V.M.; Taylor, R.; Petculescu, G. Synthesis of iron-based bulk metallic glasses as nonferromagnetic amorphous steel alloys. *Appl. Phys. Lett.* **2003**, *83*, 1131–1133. [CrossRef]
22. Inoue, A.; Nishiyama, N.; Amiya, K.; Zhang, T.; Masumoto, T. Ti-based amorphous alloys with a wide supercooled liquid region. *Mater. Lett.* **2007**, *61*, 2851–2854. [CrossRef]
23. Kim, Y.C.; Kim, W.T.; Kim, D.H. A development of Ti-based bulk metallic glass. *Mater. Sci. Eng. A* **2004**, *375*, 127–135. [CrossRef]
24. Ma, H.; Xu, J.; Ma, E. Mg-based bulk metallic glass composites with plasticity and high strength. *Appl. Phys. Lett.* **2003**, *83*, 2793–2795. [CrossRef]
25. Inoue, A.; Kato, A.; Zhang, T.; Kim, S.G.; Masumoto, T. Mg–Cu–Y Amorphous alloys with high mechanical strengths produced by a metallic mold casting method. *Mater. Trans. JIM* **1991**, *32*, 609–616. [CrossRef]
26. Choi-Yim, H.; Xu, D.H.; Johnson, W.L. Ni-based bulk metallic glass formation in the Ni-Nb-Sn and Ni-Nb-Sn-X (X = B,Fe,Cu) alloy systems. *Appl. Phys. Lett.* **2003**, *82*, 1030–1032. [CrossRef]
27. Yi, S.; Park, T.G.; Kim, D.H. Ni-based bulk amorphous alloys in the Ni-Ti-Zr-(Si, Sn) system. *J. Mater. Res.* **2000**, *15*, 2425–2430. [CrossRef]
28. Jiao, W.; Zhao, K.; Xi, X.K.; Zhao, D.Q.; Pan, M.X.; Wang, W.H. Zinc-based bulk metallic glasses. *J. Non-Cryst. Solids* **2010**, *356*, 1867–1870. [CrossRef]
29. Ge, Y.Q.; Chen, X.; Chang, Z.X. The forming and crystallization behaviors of $Zr_{50}Ti_{5}Cu_{27}Ni_{10}Al_{8}$ bulk amorphous alloy by laser additive manufacturing. *Mater. Express* **2020**, *10*, 1155–1160. [CrossRef]
30. Luo, Y.; Xing, L.L.; Jiang, Y.D.; Li, R.W.; Lu, C.; Zeng, R.G.; Luo, J.R.; Zhang, P.C.; Liu, W. Additive Manufactured Large Zr-Based Bulk Metallic Glass Composites with Desired Deformation Ability and Corrosion Resistance. *Meterials* **2020**, *13*, 597. [CrossRef]
31. Li, N.; Chen, W.; Liu, L. Thermoplastic Micro-Forming of Bulk Metallic Glasses: A Review. *JOM* **2016**, *68*, 1246–1261. [CrossRef]
32. Kim, J.; Lee, D.; Shin, S.; Lee, C. Phase evolution in $Cu_{54}Ni_{6}Zr_{22}Ti_{18}$ bulk glass Nd: YAG laser weld. *Mater. Sci. Eng. A* **2006**, *434*, 194–201. [CrossRef]
33. Kim, J.H.; Lee, C.; Lee, D.M.; Sun, J.H.; Shin, S.Y.; Bae, J.C. Pulsed Nd: YAG laser welding of $Cu_{54}Ni_{6}Zr_{22}Ti_{18}$ bulk metallic glass. *Mater. Sci. Eng. A* **2007**, *449*, 872–875. [CrossRef]
34. Louzguine-Luzgin, D.V.; Xie, G.Q.; Tsumura, T.; Fukuda, H.; Nakata, K.; Kimura, H.M.; Inoue, A. Structural investigation of Ni–Nb–Ti–Zr–Co–Cu glassy samples prepared by different welding techniques. *Mater. Sci. Eng. B* **2008**, *148*, 88–91. [CrossRef]
35. Kawahito, Y.; Terajima, T.; Kimura, H.; Kuroda, T.; Nakata, K.; Katayama, S.; Inoue, A. High-power fiber laser welding and its application to metallic glass $Zr_{55}Al_{10}Ni_{5}Cu_{30}$. *Mater. Sci. Eng. B* **2008**, *148*, 105–109. [CrossRef]
36. Wang, H.S.; Chen, H.G.; Jang, J.S.C. Microstructure evolution in Nd:YAG laser-welded $(Zr_{53}Cu_{30}Ni_{9}Al_{8})Si_{0.5}$ bulk metallic glass alloy. *J. Alloys Compd.* **2010**, *495*, 224–228. [CrossRef]
37. Wang, H.S.; Chiou, M.S.; Chen, H.G.; Jang, J.S.C.; Gu, J.W. Microstructure evolution of the laser spot welded Ni-free Zr-based bulk metallic glass composites. *Intermetallics* **2012**, *29*, 92–98. [CrossRef]
38. Pilarczyk, W.; Kania, A.; Babilas, R.; Pilarczyk, A. Structure and chosen nanomechanical properties of amorphous-crystalline laser weld. *Acta Phys. Pol. A* **2018**, *133*, 219–221. [CrossRef]
39. Wang, G.; Huang, Y.J.; Shagiev, M.; Shen, J. Laser welding of $Ti_{40}Zr_{25}Ni_{3}Cu_{12}Be_{20}$ bulk metallic glass. *Mater. Sci. Eng. A* **2012**, *541*, 33–37. [CrossRef]

40. Chen, M.H.; Lin, S.B.; Xin, L.J.; Zhou, Q.; Li, C.B.; Liu, L.; Wu, F.F. Microstructures of the pulsed laser welded TiZrBeCuMo composite amorphous alloy joint. *Opt. Lasers Eng.* **2020**, *134*, 106262. [CrossRef]
41. Shao, L.; Datye, A.; Huang, J.K.; Ketkaew, J.; Sohn, S.W.; Zhao, S.F.; Wu, S.J.; Zhang, Y.M.; Schwarz, U.D.; Schroers, J. Pulsed laser beam welding of $Pd_{43}Cu_{27}Ni_{10}P_{20}$ bulk metallic glass. *Sci. Rep.* **2017**, *7*, 7989. [CrossRef] [PubMed]
42. Pilarczyk, W.; Starczewska, O.; Lukowiec, D. Nanoindentation characteristic of Fe-based bulk metallic glass laser weld. *Phys. Status Solidi B* **2015**, *252*, 2598–2601. [CrossRef]
43. Pilarczyk, W.; Kania, A.; Pilarczyk, A. Welding and characterization of bulk metallic glasses welds. *Acta Phys. Pol. A* **2019**, *135*, 249–251. [CrossRef]
44. Wang, H.S.; Chen, H.G.; Jang, J.S.C.; Chiou, M.S. Combination of a Nd:YAG laser and a liquid cooling device to $(Zr_{53}Cu_{30}Ni_9Al_8)Si_{0.5}$ bulk metallic glass welding. *Mater. Sci. Eng. A* **2010**, *528*, 338–341. [CrossRef]
45. Wang, H.S.; Chiou, M.S.; Chen, H.G.; Jang, J.S.C. The effects of initial welding temperature and welding parameters on the crystallization behaviors of laser spot welded Zr-based bulk metallic glass. *Mater. Chem. Phys.* **2011**, *129*, 547–552. [CrossRef]
46. Wang, H.S.; Li, T.H.; Chen, H.G.; Pan, J.H.; Jang, J.S.C. Microstructural evolution and properties of laser spot-welded Zr-Al-Co-Ta bulk metallic glass under various initial welding temperatures. *Intermetallics* **2019**, *108*, 39–44. [CrossRef]
47. Chen, B.; Shi, T.L.; Li, M.; Yang, F.; Yan, F.; Liao, G.L. Laser welding of annealed $Zr_{55}Cu_{30}Ni_5Al_{10}$ bulk metallic glass. *Intermetallics* **2014**, *46*, 111–117. [CrossRef]
48. Pilarczyk, W. The investigation of the structure of bulk metallic glasses before and after laser welding. *Cryst. Res. Technol.* **2015**, *50*, 700–704. [CrossRef]
49. Wang, H.S.; Wu, J.Y.; Liu, Y.T. Effect of the volume fraction of the ex-situ reinforced Ta additions on the microstructure and properties of laser-welded Zr-based bulk metallic glass composites. *Intermetallics* **2016**, *68*, 87–94. [CrossRef]
50. Shen, Y.Y.; Li, Y.Q.; Tsai, H.L. Effect of pre-existing nuclei on crystallization during laser welding of Zr-based metallic glass. *J. Non-Cryst. Solids* **2019**, *513*, 55–63. [CrossRef]
51. Chen, B.; Shi, T.L.; Liao, G.L. Laser welding of $Zr_{41}Ti_{14}Cu_{12}Ni_{10}Be_{23}$ bulk metallic glass and zirconium metal. *J. Wuhan Univ. Technol.* **2014**, *29*, 786–788. [CrossRef]
52. Alavi, S.H.; Vora, H.D.; Dahotre, N.B.; Harimkar, S.P. Laser joining of plain carbon steel using Fe-based amorphous alloy filler powder. *J. Mater. Process. Technol.* **2016**, *238*, 55–64. [CrossRef]
53. Tao, P.J.; Zhang, W.W.; Tu, Q.; Yang, Y.Z. The Evolution of Microstructures and the Properties of Bulk Metallic Glass with Consubstantial Composition LaserWelding. *Metals* **2016**, *6*, 233. [CrossRef]
54. Wang, X.; Luo, Y.P.; Huang, T.; Liu, H.X. Experimental Investigation on Laser Impact Welding of Fe-Based Amorphous Alloys to Crystalline Copper. *Materials* **2017**, *10*, 523. [CrossRef] [PubMed]
55. Yu, X.Q.; Huang, J.K.; Shao, L.; Zhang, Y.M.; Fan, D.; Wang, Z.Y.; Yang, F.Q. Microstructures and microhardness of the welding joint between $Zr_{44}Ti_{11}Ni_{10}Cu_{10}Be_{25}$ bulk metallic glass and 1100 aluminum. *Mater. Res. Express* **2018**, *5*, 015203. [CrossRef]
56. Yu, X.Q.; Huang, J.K.; Shao, L.; Zhang, Y.M.; Fan, D.; Kang, Y.T.; Yang, F.Q. Microstructures in the joint of zirconium-based bulk metallic glass and copper. *Mater. Res. Express* **2019**, *6*, 026511. [CrossRef]

Publisher's Note: MDPI stays neutral with regard to jurisdictional claims in published maps and institutional affiliations.

© 2020 by the authors. Licensee MDPI, Basel, Switzerland. This article is an open access article distributed under the terms and conditions of the Creative Commons Attribution (CC BY) license (http://creativecommons.org/licenses/by/4.0/).

Review

Research Status and Progress of Welding Technologies for Molybdenum and Molybdenum Alloys

Qi Zhu [1], Miaoxia Xie [2,*], Xiangtao Shang [2], Geng An [1,3], Jun Sun [3], Na Wang [1], Sha Xi [1], Chunyang Bu [1] and Juping Zhang [1]

[1] Technical Center, Jinduicheng Molybdenum Co., Ltd., Xi'an 710077, China; zhuyaqian2009@163.com (Q.Z.); gmail@163.com (G.A.); biyewangna@163.com (N.W.); xs19861105@126.com (S.X.); bp208@163.com (C.B.); hjsjdc@yeah.net (J.Z.)
[2] School of Mechanical and Electrical Engineering, Xi'an University of Architecture and Technology, Xi'an 710055, China; shang_x_t@163.com
[3] State key laboratory for mechanical behavior of materials, Xi'an Jiaotong University, Xi'an 710049, China; junsun@mail.xjtu.edu.cn
* Correspondence: Xiemiaoxia@xauat.edu.cn; Tel.: +86-181-4902-5125

Received: 28 December 2019; Accepted: 14 February 2020; Published: 20 February 2020

Abstract: Owing to its potential application prospect in novel accident tolerant fuel, molybdenum alloys and their welding technologies have gained great importance in recent years. The challenges of welding molybdenum alloys come from two aspects: one is related to its powder metallurgy manufacturing process, and the other is its inherent characteristics of refractory metal. The welding of powder metallurgy materials has been associated with issues such as porosity, contamination, and inclusions, at levels which tend to degrade the service performances of a welded joint. Refractory metals usually present poor weldability due to embrittlement of the fusion zone as a result of impurities segregation and the grain coarsening in the heat-affected zone. A critical review of the current state of the art of welding Mo alloys components is presented. The advantages and disadvantages of the various methods, i.e., electron-beam welding (EBW), tungsten-arc inert gas (TIG) welding, laser welding (LW), electric resistance welding (ERW), and brazing and friction welding (FW) in joining Mo and Mo alloys, are discussed with a view to imagine future directions. This review suggests that more attention should be paid to high energy density laser welding and the mechanism and technology of welding Mo alloys under hyperbaric environment.

Keywords: molybdenum alloy; welding; status; progress

1. Introduction

Molybdenum (Mo) and Mo alloys show characteristics such as high melting point, good high-temperature strength, high wear resistance, high thermal conductivity and low resistivity, low coefficient of linear expansion, high elastic modulus, and good corrosion resistance [1]. Based on this, they have irreplaceable functions and application demands in the fields like the defense industry, aerospace, electronic information, energy, chemical defense, metallurgy, and nuclear industry. However, Mo and Mo alloys are hard and brittle materials in nature, so their weldabilities are generally poor [2]. There are two main sources of molybdenum brittleness: one is the intrinsic brittleness of molybdenum, and the other is the enrichment of interstitial impurities in the grain boundary. Oxygen is the most important impurity element in the grain boundary which affects the embrittlement of molybdenum. The solubility of oxygen in molybdenum at room temperature is less than 0.1 ppm, which forms relatively volatile molybdenum oxide at the grain boundary, which greatly reduces the bond strength of grain boundary. After melting and welding of high-performance molybdenum alloy, the weld forms an as-cast structure with coarse grains, the heat-affected zone forms a coarse recrystallization structure, the impurity elements are enriched in the grain boundary, and the strength and toughness of the weld and heat-affected zone are greatly reduced [3–5]. To extend application field of Mo and Mo alloys, researchers worldwide have conducted a lot of studies on their welding and relevant kinds of literature have been reported since the 1970s [1].

In recent years, the global nuclear industry and scientific community have been aware that a new fuel system, that is, accident tolerant fuel (ATF) needs to be developed [6,7]. Such a fuel system needs to be able to withstand severe accident conditions and slow down the rate of deterioration over a long period of time, to provide more valuable time for people to take emergency measures and greatly reduce the risks of leakage of radioactive materials. Therefore, Mo alloy is listed as one of the main candidate materials for ATF cladding by the global nuclear industry [8]. In this context, the welding technologies for Mo and Mo alloys have attracted wide attention of researchers in China and a lot of new progress has been achieved in recent years.

2. Analysis on Weldability of Mo and Mo Alloy

2.1. Room-Temperature Brittleness

Ductility of most of Mo alloys varies with temperature to make the materials change from ductile fracture to brittle fracture in a very small temperature range. The ductile-brittle transition temperature of pure Mo ranges from approximately 140 °C to 150 °C, resulting in difficulties in intensive processing, low product performance, and limited application fields [1]. Such brittleness is known as intrinsic brittleness of Mo, which is mainly determined by an electron distribution characteristic that the outermost and sub-outermost electrons of its atoms are half full.

Mo and Mo alloy have a high melting point, good thermal conductivity, high recrystallization temperature, no allotropy transformations in solid state, and low density of bcc crystal structure. Due to these characteristics, weld seam and heat-affected zone (HAZ) is large and grains are seriously coarsened after welding (Figure 1), so that interstitial impurities, such as C, N, and O are fully diffused and enriched on grain boundaries, resulting in greatly weakened bonding strength of grain boundaries, like Figure 2 that the fracture of laser welded Mo alloy contains a lot of MoO_2. Under the joint effects of intrinsic brittleness of the materials and segregation of impurities at grain boundaries, sensitivity of welding cracks is high and strength, plasticity, and ductility of Mo and Mo alloy joints are poor [9,10]. Therefore, molybdenum and molybdenum alloy parts or structures are usually manufactured by powder metallurgy rather than welding.

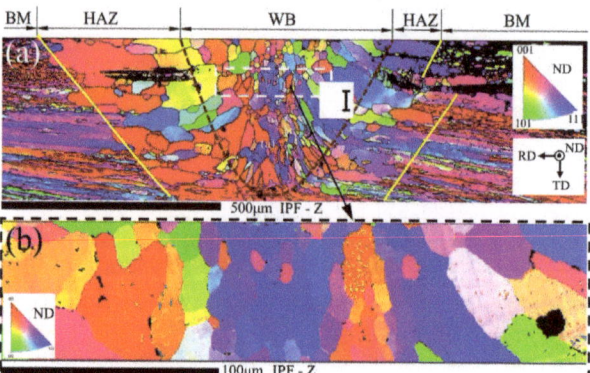

Figure 1. (a) Electron backscatter diffraction (EBSD) image of cross section of Mo joint achieved by laser welding, and (b) enlarged view of area (a) [10].

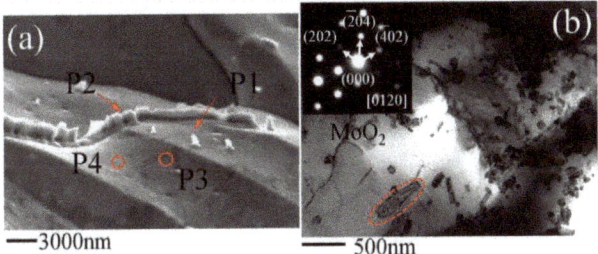

Figure 2. (a) SEM image of the fracture of Mo joint with plenty of MoO$_2$ at grain boundary, and (b) TEM bright field image of the weld bead of the Mo joint [10].

2.2. Pore Defects

Furthermore, owing to the powder metallurgy process that can yield fine grain structure without preferred orientation, the refractory metal work blanks are usually prepared by utilizing powder metallurgy method. This leads to that the material contains micropores and impurity elements, and its compactness is not comparable to that of the materials produced by smelting metallurgy. Therefore, the problem of high rate of pore defects (Figure 3) generally appears in fusion-welded Mo and Mo alloys. Particularly, the high-pressure residual gases in the micropores are the most harmful. During the welding process, these high-pressure gases can expand rapidly in the molten pool after being released into the high-temperature molten pool, which seriously deteriorates the quality of welded joints of Mo and Mo alloys [11–13].

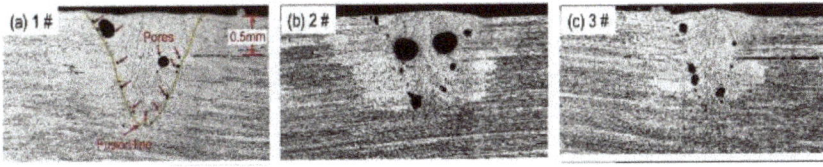

Figure 3. Typical porosity morphology in Mo joints produced by laser welding: (a) orthogonal experiment 1# parameter sample, (b) orthogonal experiment 2# parameter sample, and (c) orthogonal experiment 3# parameter sample. [11].

3. Research Progress in Welding of Mo and Mo Alloys

At present, the welding methods for Mo and Mo alloys mainly include tungsten-arc inert gas (TIG) welding, electron-beam welding (EBW), laser welding, electric resistance welding (ERW), brazing, and friction welding.

3.1. EBW

Pan et al. [14] studied EBW for pure Mo with a thickness of 1.5 mm obtained by powder metallurgy. The results show that the faster the welding speed is, the small the grain size and the less the interstitial impurities. By increasing welding speed and reducing welding heat input, the ductility of welded joint of Mo can be significantly improved. Vacuum degree significantly affects ductile-brittle transition temperature, whereas the decomposition of oxides on the surface of the workpiece during welding has great influence on the vacuum degree. When a vacuum degree increases from 10–4 mm Hg to 10–5 mm Hg, the upper limit of ductile-brittle transition temperature of the welded joint decreases from ~150 °C to ~100 °C. Pan et al.'s analysis of the change in joint performance through the impurity composition and impurity content of the gas released after the melting of the material left a deep impression on the authors. Yang et al. [15] welded pure Mo plates with thickness of 16 mm by using

EBW. The results demonstrate that the highest strength of the welded joint subjected to heat treatment at 1100 °C was found in the weld seam. Tensile fractures of the welded joint are shown in the weld seam and present morphology of cleavage fracture. In addition, Zheng et al. [16] welded pure Mo materials with thickness of 16 mm through vacuum EBW. As can be seen from the above research results, the results illustrate that grains in weld seam of Mo by EBW grow rapidly.

Morito et al. (1989) [17] found that a joint of molybdenum-titanium-zirconium (TZM) alloys welded using EBW at room temperature always shows brittle fracture. However, at temperature higher than 300 °C, the joint always presents ductile fracture, and there is obvious necking phenomenon before fracturing. In addition, the research demonstrates that carbonization and heat treatment after welding can effectively raise the strength of the joint of Mo alloys, which is primarily attributed to an increase of the grain boundary cohesion due to the effective carbon segregation and precipitation. [18]. Morito et al. (1998) [19] found that the strength and ductility of a welded joint of Mo alloys obtained through EBW can rise after increasing rhenium (Re) content. The reason is that two-phase microstructure is formed in the weld zone with enhanced Re content. Based on the thermal simulation test, Morito et al. (1997) [20] compared ductilities of HAZs when welding Mo alloys (Mo > 99.9 wt%) under two conditions of thermal treatment, i.e., furnace cooling and quick cooling through quenching. It is found that quick cooling after welding can significantly reduce the ductility of HAZ of the welded joint of Mo alloys, mainly because grain-boundary segregation in HAZ is more significant under quick cooling through quenching. Stütz et al. (2016) [21] systematically studied the influences of parameters of EBW process on sizes of fusion zone (FZ) and HAZ, size of grains in FZ and HAZ, sensitivity to pores and cracks in butt welded joint of TZM alloy with thickness of 2 mm. The results show that pore defects are serious when welding heat input is large. Small heat input can not only inhibit pores, but also obviously reduce grain size in FZ. The strength of the joint welded by EBW can reach 50–77%: that of the base metal (BM). Therefore, Stütz et al. pointed out that it was necessary to study EBW in terms of filling materials and alloying metals in the weld seam, which has important guiding significance for improving mechanical properties of molybdenum alloy joints. Recently, Chen et al. (2018) [22] conducted electron beam welding of molybdenum and found that the even tensile strength of the joints was 280 MPa, and the fracture position was located in the weld, which was a brittle fracture, the fracture location is shown in Figure 4. It was determined as quasi-cleavage fracture. Pore and crack defects were observed in the weld zone. The pores were formed by the oxygen that was not escaping the molten pool. Cracks were confirmed as solidification cracks and low plastic embrittlement cracks.

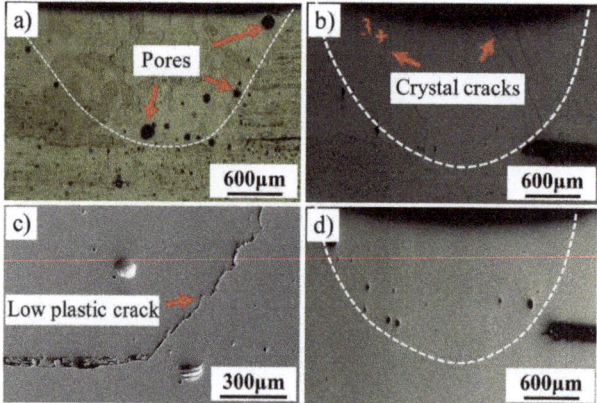

Figure 4. Cross-sectional morphology of the welded joints produced by electron-beam welding (EBW) welding (**a**) pore defects diagram with improved parameters, (**b**) crystal cracks, (**c**) low plastic embrittlement crack, and (**d**) forming after optimized process. [22].

In addition, Chen et al. (2019) [23] found that during welding with a 0.6 mm beam deflection to Kovar, the weld zone exhibits equiaxed crystals. Compared with the columnar crystal during welding without beam deflection, the microstructure of the weld zone transforms into an equiaxed crystal when welding with a 0.6 mm beam offset to Kovar. The morphology of the reaction layer alters due to the beam deflection, which escalates its toughness. No cracks are observed in the heat-affected zone on the molybdenum side. The tensile strength of the joint increases resulting from the beam deflection to Kovar, which exceeds 260 MPa when the beam deflection is 0.6 mm.

3.2. TIG Welding

Wang et al. [24] studied the TIG welding of TZM alloy. The results demonstrate that well-formed weld seam can be obtained under proper welding current (as is shown in Figure 5), welding speed, and argon gas flow, better welding process parameters: welding speed 4 mm/s, argon flow rate 10 L/min, welding current should be controlled at ~210 A. Jiang et al. [25] researched TIG welding of Mo-Cu composite materials and stainless steel filled with Cr-Ni wires.

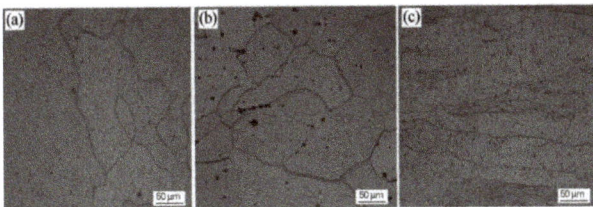

Figure 5. Microstructure of weld joint (**a**) welding seam zone, (**b**) heat-affected zone, and (**c**) TZM martix. [24].

Wang [26] found that there are fewer pores in weld seams obtained by EBW or TIG welding of smelting Mo alloy, whereas there are more pores in weld seams of pure Mo or Mo alloys obtained by powder metallurgy. The addition of C can improve plasticity of powder-metallurgy-processed weld seam of pure Mo or Mo alloys by reducing content of molybdenum oxide at grain boundary, and significantly reduce pores in the weld seam. In addition, adding Ti and hafnium (Hf) into the weld seam can decrease centerline cracks and pores in weld seam of powder-metallurgy-processed pure Mo, so that tensile fracture transfers from weld seam to HAZ.

Matsuda et al. [27] studied EBW and TIG welding of TZM alloy with thickness of a 1.5 mm prepared through powder metallurgy. The research shows that large welding heat input can significantly decrease ductility of a welded TZM alloy joint and the ductile-brittle transition temperature of the welded joint obtained through TIG welding is ~120 °C higher than that of EBW. Furthermore, they also found that pore defects only appear around the arc starting and arc extinguishing positions in weld seam during TIG welding, whereas pore defects greatly increase in weld seam during EBW welding in the vacuum environment. Kolarikova et al. (2012) [28] investigated EBW and TIG welding of pure Mo sheets. Widths of FZs in joints obtained through EBW and gas tungsten arc welding (GTAW) separately are 0.8 mm and 1.7 mm, whereas HAZs are significantly different in width (1.4 mm and 35 mm). Grain sizes in FZ and HAZ in the EBW joint are obviously smaller than those in the GTAW joint, indicating that EBW with high energy density is more suitable for welding Mo than GTAW

3.3. Laser Welding

Liu et al. (2016) [29] studied continuous-wave Nd:yttrium-aluminum-garnet (YAG) laser welding of an overlap joint of Mo-Re alloy (50Mo:50Re) with the thickness of 0.13 mm prepared by powder metallurgy. After welding, cracks appear at the bonding interface of FZ, and many large pores are observed at the bonding interface of FZ, with the diameter being ~15–20% of the thickness of BM plates. Microscopic analysis results of fracture present that intergranular fracture occurs and there are a large number of dark compounds in the grains and on the grain boundary, as is shown in Figure 6, and the composition analysis showed that the content of C and O in these compounds were 30% and 15%, respectively. It is believed that coarse microstructures and harmful impurity elements cause the hardening of bonding interface and intergranular fracturing of the joint. The study of Lin (2013) [3] demonstrates that welding conductive elements of needle-shaped pure Mo with a diameter of 0.5 mm using pulse Nd:YAG laser welding instead of ERW can significantly raise the strength of the joint.

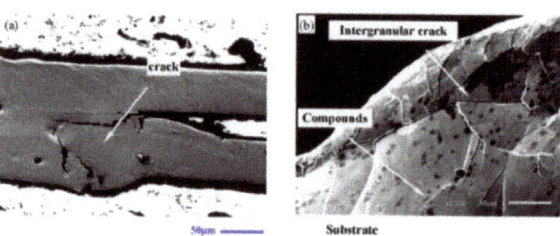

Figure 6. SEM observation of 50Mo-50Re overlap joint achieved by laser welding: (**a**) cross section and and (**b**) fracture surface [29].

Kramer et al. (2013) [4] studied EBW and pulse Nd:YAG laser welding of Mo-44.5% Re alloy sheets with a thickness of 0.5 mm. The research shows that a Mo-44.5% Re alloy joint welded by EBW is well-formed, without defects including pores and cracks. Cracks are found in FZ in the laser-welded Mo-44.5% Re alloy joint and the micromorphology of brittle fracture is shown in the fracture of the laser-welded joint after testing mechanical performances. Chatterjee et al. (2016) [5] researched EBW and Nd:YAG laser-TIG hybrid welding for butt welded joint of wrought TAM alloy (Ti 0.50 wt%, Zr 0.08 wt% and C 0.04 wt%) with thickness of 1.2 mm. Grain sizes in FZ and HAZ in the EBW joint are obviously small, which are ~55% and ~65% of those in the joint obtained by Nd:YAG laser-TIG hybrid welding. The weld width of EBW and hybrid welding method is ~1.4 mm and 2.6 mm respectively, but in both cases, the width of HAZ is ~1.5 times of the weld width. Results of tensile test demonstrate that strengths of joints prepared by Nd:YAG laser-TIG hybrid welding and EBW are about 41% and 47% that of BM. The fracture morphology is shown in Figure 7, the two joints hardly show any tensile plasticity in the tensile test, and the shrinkage and elongation of cross sections are almost zero, while the elongation of BM is up to 8.4%. Figure 8 shows the presence of large grains and nearly uniform distribution of a second phase within the gain, typical volume fraction estimated from these micrographs shows that the oxide phase is nearly 10 pct of the volume fraction.

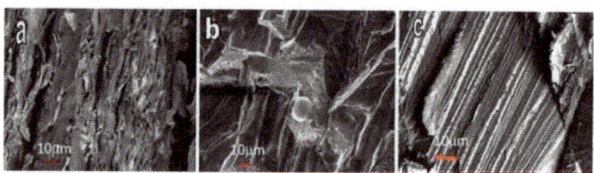

Figure 7. Fractographs of (**a**) the parent metal, (**b**) EB weld joint, and (**c**) LGTHW Joint of tensile samples. Change in the morphology and the presence of sharp faceting in samples containing weld joints could be noticed [5].

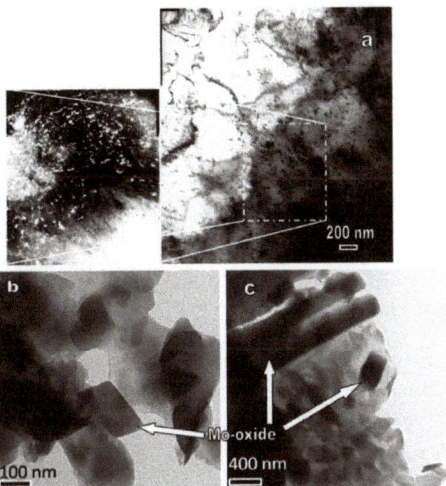

Figure 8. TEM micrographs showing (**a**) the presence of the precipitates within the matrix. Inset shows the magnified view; (**b**) Magnified view showing the presence of Mo-oxide phase in the weld region. (**c**) The presence of needle-shaped long oxides may be noticed [5].

An et al. (2018) [30] carried out laser lap welding of fuel cladding and end plug made of molybdenum (Mo) alloy. Under the optimum processing conditions, the tensile strength of the welded joint reached 617 MPa, taking up 82.3% that of the base metal. Recently, Zhang et al. (2019) [9] successfully enhanced the mechanical performance of fusion zone in laser beam welding joint of molybdenum alloy by solid carburizing. As is shown in Figure 9 the tensile strength of carburized weld joints rose by 426% compared with that of uncarburized weld joints. The TEM images of molybdenum oxide particles in the FZ of LW joint show that the particles were shown as lenticular or elongated blocks under TEM observation (Figure 10a) and judged from the diffraction pattern as MoO_2. And the TEM images of molybdenum carbide particles in the FZ of SC-150 joint show that the particles were shown as circular pattern under TEM observation (Figure 10b) and judged from the diffraction pattern as Mo_2C.

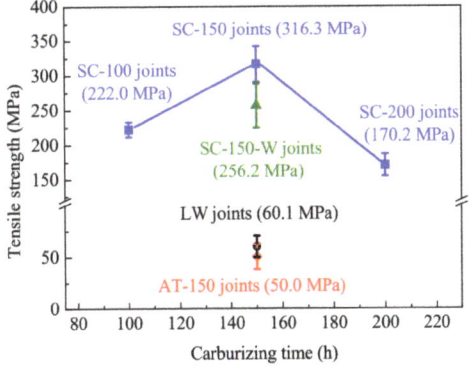

Figure 9. The effect of C addition on tensile strength of the Mo alloy joints [9].

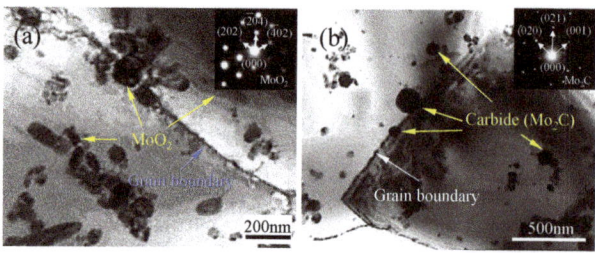

Figure 10. TEM images of (**a**) molybdenum oxide particles in the FZ of LW joint and (**b**) carbide particles in the FZ of SC-150 joint [9].

Xie et al. (2019) [12] found that low heat input (i.e., high welding speed) resulted in significantly refined grains in the fusion zone (FZ) of fiber laser welded nano-sized rare earth oxide particles and superfine crystal microstructure (NS) Mo joints, the cross-sectional microstructures of the NS Mo alloy laser welding joints is shown in Figure 11. When welding heat input decreased from 3600 J/cm (i.e., 1.2 kW, 20 cm/min) to 250 J/cm (i.e., 2.5 kW, 600 cm/min), the tensile strength of welded joints increased from ~250 MPa to ~570 MPa. They also found that laser welding of NS-Mo under low heat input significantly reduced the porosity defects in the fusion zone (2019) [13].

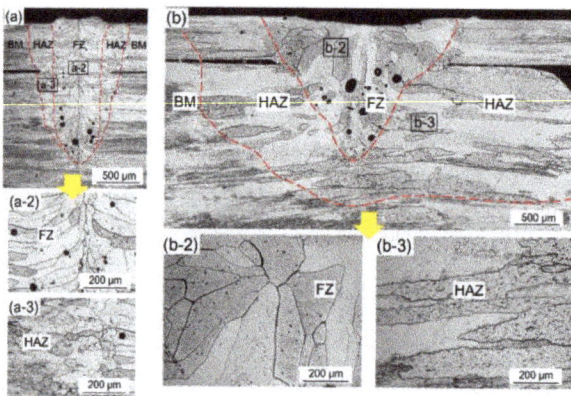

Figure 11. The cross-sectional microstructures of the NS Mo alloy laser welding joints. Heat input: (**a**) 250 J/cm and (**b**) 3600 J/cm [12].

In the study of Zhang et al. (2019) [10], titanium was selected as an alloying element to reduce brittleness of laser weld beads in molybdenum "cladding-end plug" socket joints. Brazing was also performed to enhance joint strength. The combined structure of laser welding and brazing is shown in Figure 12, joints with the same strength as base material and hydraulic bursting pressure of 60 MPa were produced using a combination of the two methods.

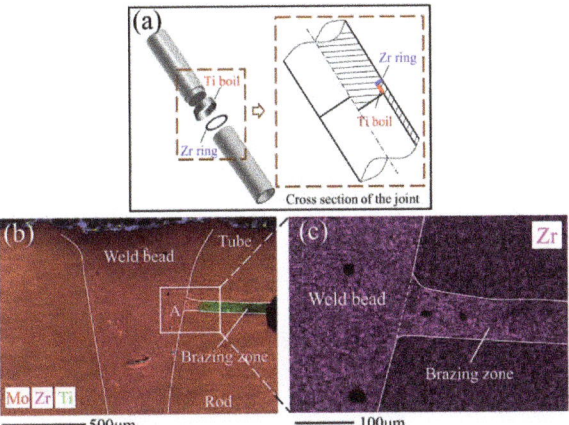

Figure 12. Tracing of the metal elements in the weld bead of the Mo-0.03Ti-B joint: (**a**) schematic diagram of the location where Zr was added. (**b**) Analysis of Mo, Zr, and Ti on the cross section of the joint by EDS surface scanning. (**c**) Analysis of Zr in zone A in panel (**b**) by EDS surface scanning. [10].

Liu et al. (2019) [11] found that welding cycles had a significant influence on the porosity ratio of fusion zone (FZ), whereas the amplitude and frequency of laser power waveform slightly influenced the porosity. Moreover, Zr added in a molten pool can be preferentially reacted with O to generate ZrO_2, which can inhibit the precipitation of volatile MoO_2 to thus suppress the generation of metallurgy-induced pores, reconstructed 3D transparent distribution of pores in joints is shown in Figure 13. Zhang et al. (2019) [7] conducted laser seal welding of end plug to thin-walled nanostructured high-strength molybdenum alloy cladding with a zirconium interlayer and tensile strength of the achieved welded joints matched that of the base metal. Note that by taking advantage of the metallurgical characteristics of molybdenum and its high melting point and high thermal conductivity, Zhang et al. [9–11] put forward a systematic strategy that can effectively solve the problems of porosity and embrittlement in welding fuel cladding made of molybdenum alloys. More importantly, the results of tensile test and hydraulic bursting show that the molybdenum alloy cladding prepared by this method have excellent performance, which completely eliminates the doubts about the welding quality of molybdenum alloy fuel cladding, and is of great significance to the promotion and application of molybdenum alloy accident-tolerant fuel cladding, the results of tensile test and hydrostatic test are shown in Figure 14.

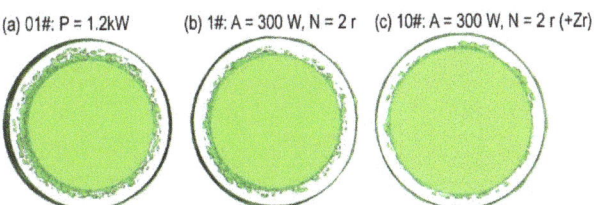

Figure 13. Reconstructed 3D transparent distribution of pores in the three joints achieved under (**a**) P = 1.2 kW; (**b**) average power = 1.2 kW, Amplitude = 300 W, frequency = 50 HZ, N = 2 cycles; and (**c**) average power = 1.2 kW, amplitude = 300 W, frequency = 150 HZ, N = 2 cycles, adding Zr, respectively. [11].

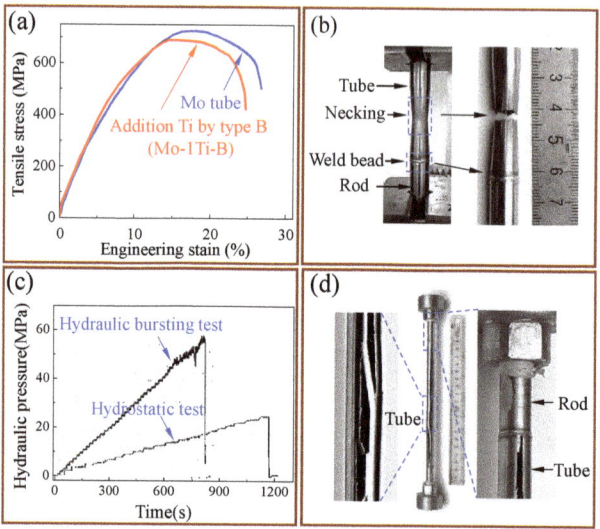

Figure 14. (a) Strength-displacement curves of Mo-0.03Ti-B and pure Mo tube; (b) images of the joint after fracture testing; (c) hydrostatic test curve and hydraulic bursting test curve for the molybdenum joint of Mo-0.03Ti-B; and (d) images of joint after hydraulic bursting testing [10].

In the study of Wang et al. (2019) [31], laser beam offset welding was used to join pure Mo and 304L. The results demonstrated that tensile strength of the joints could be increased to ~280 MPa by presetting a nickel (Ni) foil at Mo/304 L interface and shifting laser beam to 304L, whereas the tensile strength of the sample without Ni foil is only 112 MPa, and the results of tensile test is shown in Figure 15. In the study of Lu et al. (2018) [32,33], by adding zirconium (Zr) to the molten pool, ultimate tensile strength (UTS) of the dissimilar joint of titanium and molybdenum was increased from about 350 MPa to ~470 MPa, which reached more than 90% of that of the Ti base metal (BM). Zhou et al. (2018) [34] observed cracks in dissimilar laser welding of tantalum to molybdenum and pointed out that solidification cracking tendency of Mo was the main reason for crack initiation in the Ta/Mo joint. Ning et al. (2019) [35] studied the potential of laser welding of 0.5 mm-thick Titanium-zirconium-molybdenum (TZM) alloy in a lap welding configuration. They found that introducing an interface gap of 0.09 mm had the most positive effect in reducing the porosity compared to using helium gas, different shielding gas flow rates, adding alloy elements, and different heat input rates. Liu et al. (2019) [36] compared the micro-structures, properties, and residual stresses of the welded girth joints achieved at different preheating temperatures and found that the tensile strength reached a maximum at the preheating temperature of 673 K, which was approximately 50% that of the base metal. Gao et al. (2020) [37] also studied the effect of laser offset on microstructure and mechanical properties of laser welding of pure molybdenum to stainless steel. As the laser beam shifts from the Mo side to the stainless steel side, the formation of welding defects and Fe-Mo intermetallic compounds (IMCs) are effectively restricted because of the decrease amount of molten Mo. Consequently, the tensile strength of joints increased first and then decreased in the laser offset range of 0.2–0.5 mm. The highest tensile strength of the joints is 290 MPa at the laser offset of 0.3 mm.

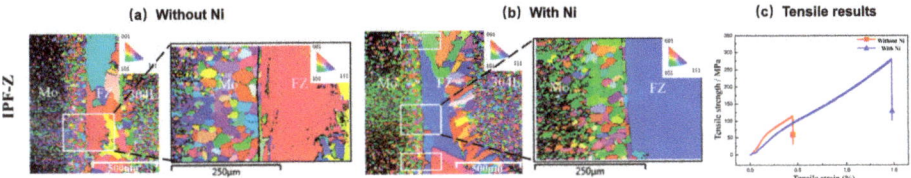

Figure 15. (a) Fracture cross section of the joint without Ni, (b) fracture cross section of the joint with Ni, and (c) tensile results of the two joints [31].

3.4. ERW

Xu et al. (2007) [38] investigated the optimization of the resistance spot welding process of an overlap joint of 50Mo-50Re (wt%) alloy with thickness of 0.127 mm. The study found that the longer the time of application of upsetting force after power off is, the higher the strength of the joint and the better the ductility. When time of applying upsetting force after power off increases from 50 ms to 999 ms, bearing capacity of the joint rises from 100 N to 113 N and fracture mode changes from brittle intergranular fracture to dimple fracture. This is because after power off, the increase of time of applying upsetting force can accelerate the cooling rate of weld seam, thus inhibiting segregation of Mo at grain boundaries. It is also found from the study that the defects of large-size pores appear in FZ of the joint under various welding conditions, because there are micropores in powder metallurgy materials and is shown in Figure 16.

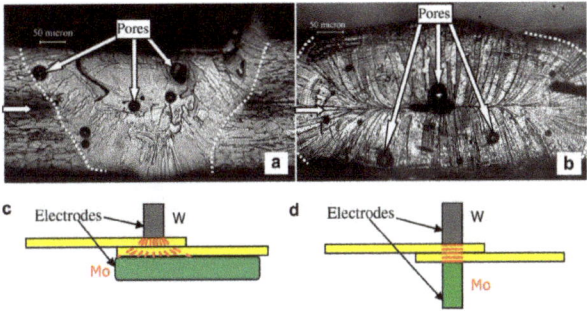

Figure 16. (a,b) The shape and microstructure of the nuggets welded using the electrodes shown in (c,d), respectively. The horizontal arrows show the interfaces between two workpieces in panels (a) and (b) [38].

Elizabeth E. Ferrenz et al. [39] controlled welding quality by using double pulse current waveforms in resistance spot welding of Mo and T-Re alloy wires. The first pulse current is small and mainly used to remove oxide film, while the second pulse is used for welding with a larger current.

3.5. Brazing

Xia et al. (2017) [40] studied vacuum brazing for an overlap joint of 50Mo-50Re alloy with a thickness of 0.06 mm. BM is prepared by utilizing powder metallurgy. The brazing filler metal is Ni-Cr-Si-B (Ni-19Cr-7.3Si-1.5B wt%) with a melting temperature range of 1081 to 1136 °C. After heat preservation for 20 min at brazing temperature of 1200 °C, the well-formed brazing seam was obtained, without defects, such as microcracks and pores. However, CrB and NiSi2 brittle intermetallic compounds are formed at the center of the brazing seam. Song et al. (2015) [41] studied vacuum brazing for overlap joint of TZM alloy (Ti 0.50 wt%, Zr 0.08 wt%, and C 0.04 wt%) with thickness of 3 mm. The brazing filler metal is Ti-28Ni (wt%) eutectic filler metal with melting temperature in the range of 940 to 980 °C. The range of brazing temperature is 1000–1160 °C and vacuum degree is

~1.33 MPa. The shear strength of the brazing joint preserved for 600 s at 1080 °C reaches ~107 MPa. The shear fracture shows the morphology of quasi-cleavage transgranular fracture.

3.6. Friction Welding

Fu et al. [42] studied friction welding of Mo alloy and die steel. The results demonstrate that thermal coupling during the friction welding process is conducive to grain refinement near seam and closure of pores in TZM powder alloy. According to strict welding specifications, a good joint without defects can be obtained through friction welding. Yazdanian et al. [43] researched friction stir welding of pure Mo plate (99.5 wt%) with thickness of 1.5 mm by utilizing a stir-welding head of Iridium (Ir)-Re alloy. The strength of butt welded joint prepared at rotation speed of 1000 rpm and welding speed of 100 mm/min reaches 86% that of BM and the joint is fractured in HAZ in a tensile test. Reheis et al. (2014) [44] investigated continuous drive friction welding of a TZM alloy tube with an outer diameter of 55 mm and wall thickness of 7.5 mm. A well-formed joint is obtained under the optimized process parameters, and its tensile strength at room-temperature is equivalent to that of BM, whereas its elongation is ~50% lower than that of BM. Ambroziak et al. (2011) [45] studied continuous drive friction welding of a refractory metal bar with the diameter of 30 mm under different combinations, such as Mo-Mo, TZM-TZM, TZM-V, TZM-Ta, Mo-Nb, and TZM-NB. In the whole welding process, the sample was immersed in IME82 oil (Figure 17) to prevent the workpiece from being polluted by environmental gas at high temperature. The results present that the well-formed welded joint with fine grains can be obtained for each combination under reasonable technological conditions.

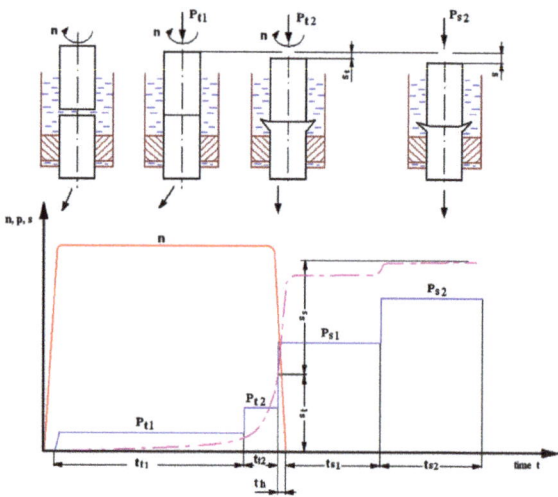

Figure 17. Scheme of friction welding in IME82 oil [45].

Recently, Stütz et al. (2016) [46] realized continuous drive friction welding of a pure Mo tube with wall thickness of 10 mm and outer diameter of 130 mm (Figure 18). In the study of Stutz et al. (2018) [47,48], continuous dynamic recrystallization and the competing dynamic recovery were observed as key mechanisms; intensive subgrain formation and the onset of recrystallization played the major role on the microstructure modification due to rotary friction welding. Grain refinement is observed in the weld interface for the TZM, whereas coarse grains are observed in the same zone for pure Mo but comparable crystallographic texture is observed for both materials, the result is shown in Figure 19.

Figure 18. Medium-size friction welded Mo-tube (OD: 150 mm, ID: 130mm) [46].

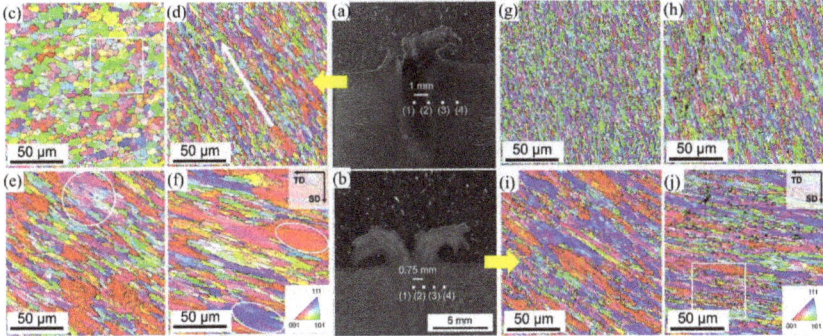

Figure 19. Detail of the microstructure friction welded samples: Overview of (**a**) Mo; overview of (**b**) TZM; (**c**) zone (1) of panel (**a**); (**d**) zone (2) of panel (**a**); (**e**) zone (3) of panel (**a**); (**f**) zone (4) of panel (**a**); (**g**) zone (1) of panel (**b**); (**h**) zone (2) of panel (**b**); (**i**) zone (3) of panel (**b**); (**j**) zone (4) of panel (**b**). [48].

4. Summary and Prospects

4.1. Advantages and Disadvantages of Various Methods for Welding Mo and Mo Alloys

Mo and Mo alloys are commonly used as high-temperature structural materials under harsh working conditions. Due to low welding temperature and uniform heating of weldment, brazing incurs small deformation and is easy to ensure the accurate size of weldment. However, strength and heat resistance of brazing seam are lower than those of BM and its performance at high temperature is generally inferior to that of joints obtained by fusion welding. High electric conductivity and yield strength at high temperature of Mo worsen its weldability in ERW. In friction welding, the instrument is worn seriously and an irreparable keyhole is formed in the weld seam when lifting the stir-welding head out from the workpiece after welding. Moreover, corrosion resistance of the weld seam reduces and it is difficult to clamp the components of the thin-walled tube. The high thermal conductivity, the significant tendency of grain coarsening, and grain-boundary embrittlement of Mo alloys determining that welding using a heat source with high power density has great advantages for Mo and Mo alloys. EBW with high energy density can be used for welding refractory alloys and difficult-to-weld materials, and shows fast welding speed, small HAZ, and small welding stress and deformation. The vacuum environment for EBW can not only prevent the molten metal from being polluted by gases, such as oxygen and nitrogen, but also facilitate the degassing and purification of metal in weld seam. Therefore, EBW is widely used in welding of Mo and Mo alloys. However, EBW has shortcomings, such as complex process, low efficiency, limited size and shape of weldment by the vacuum chamber, and susceptibility to the interference of stray electromagnetic field and X-ray radiation produced during welding. In recent years, fiber laser and Disk laser technologies with excellent beam quality have developed rapidly, providing opportunities and necessary conditions for the breakthrough and development of welding technologies for Mo and Mo alloys. Laser welding not only has many advantages afforded by heat sources with high energy beams, such as high-power density and small heat input, but also can be carried out in an open environment. In the meanwhile, it is more difficult to protect the high temperature zone and control pore defects in weld seam during laser welding of Mo and Mo alloys compared with EBW. In addition, note that the last welding spot of the nuclear fuel rod needs to be welded under hyperbaric environment to encapsulate the hyperbaric inert gas, so the mechanism and technology of laser welding molybdenum alloy under hyperbaric environment is a blank field to be studied urgently.

4.2. Demands and Prospects

In recent years, ATF fuel has gained attention in technical research and development in the international community of nuclear energy. Nuclear energy agencies in these countries, such as the United States, France, South Korea, and Japan, take ATF as their next key development direction. In 2016, Professor Stephen Zinkel, an academician of the National Academy of Engineering in the United States, pointed out that ATF fuel is profoundly influencing the development direction of science and technology of nuclear energy and will change "game rules" for nuclear safety and nuclear power industry in the world. High-performance Mo alloy is a main alternative material for the next generation of ATF cladding materials. Although it has excellent ductility, its weldability remains to be solved, so development of reliable welding technology for Mo alloy has become a pressing need. Studying problems, such as embrittlement and pore defects in welding of Mo and Mo alloys and exploring new methods and mechanisms for controlling welding quality, not only has important theoretical significance, but also shows important engineering application significance.

Author Contributions: Q.Z. and M.X. wrote the major part of the review; X.S., G.A., J.S., N.W., S.X., C.B., and J.Z. participated on the concept of review and made the corrections/suggestions for improvement. All authors have read and agreed to the published version of the manuscript.

Funding: This work was supported by the National Natural Science Foundation of China (Grant No. 51775416).

Conflicts of Interest: The authors declare no conflict of interest.

References

1. Doane, D.V.; Timmons, G.A.; Hallada, C.J. Molybdenum and Molybdenum Alloys. In *Kirk-Othmer Encyclopedia of Chemical Technology*; John Wiley Sons, Inc.: Hoboken, NJ, USA, 2000.
2. Scott, M.; Knowlson, P. The welding and brazing of the refractory metals niobium, tantalum, molybdenum and tungsten—A review. *J. Less Common Met.* **1963**, *5*, 205–244. [CrossRef]
3. Lin, Y.; Jiang, G. Pulsed Nd:YAG laser fine spot welding for attachment of refractory mini-pins. *SPIE LASE* **2013**, *8608*, 885–905.
4. Kramer, D.P.; Mcdougal, J.R.; Booher, B.A.; Ruhkamp, J.D. Electron beam and Nd-YAG laser welding of niobium-1% zirconium and molybdenum-44.5% rhenium thin select material. In Proceedings of the Energy Conversion Engineering Conference and Exhibit, Las Vegas, NV, USA, 24–28 July 2000; Volume 2, pp. 956–961.
5. Chatterjee, A.; Kumar, S.; Tewari, R.; Dey, G.K. Welding of mo-based alloy using electron beam and laser-gtaw hybrid welding techniques. *Metall. Mater. Trans. A* **2016**, *47*, 1–10. [CrossRef]
6. Ott, L.; Robb, K.; Wang, D. Corrigendum to "Preliminary assessment of accident-tolerant fuels on LWR performance during normal operation and under DB and BDB accident conditions" [J. Nucl. Mater. 448 (2014) 520–533]. *J. Nucl. Mater.* **2015**, *461*, 178–179. [CrossRef]
7. Zhang, L.-J.; Pei, J.-Y.; Zhang, L.-L.; Long, J.; Zhang, J.-X.; Na, S.-J.; Zhang, L.-J.; Pei, J.-Y.; Liang-Liang, Z. Laser seal welding of end plug to thin-walled nanostructured high-strength molybdenum alloy cladding with a zirconium interlayer. *J. Mater. Process. Technol.* **2019**, *267*, 338–347. [CrossRef]

8. Pint, B.A.; Terrani, K.A.; Yamamoto, Y.; Snead, L.L. Material Selection for Accident Tolerant Fuel Cladding. *Met. Mater. Trans. E* **2015**, *2*, 190–196. [CrossRef]
9. Zhang, L.-L.; Zhang, L.-J.; Long, J.; Sun, X.; Zhang, J.-X.; Na, S.-J. Enhanced mechanical performance of fusion zone in laser beam welding joint of molybdenum alloy due to solid carburizing. *Mater. Des.* **2019**, *181*, 107957. [CrossRef]
10. Zhang, L.-L.; Zhang, L.-J.; Long, J.; Ning, J.; Zhang, J.-X.; Na, S.-J. Effects of titanium on grain boundary strength in molybdenum laser weld bead and formation and strengthening mechanisms of brazing layer. *Mater. Des.* **2019**, *169*, 107681. [CrossRef]
11. Zhang, L.-J.; Liu, J.-Z.; Pei, J.-Y.; Ning, J.; Zhang, L.-L.; Long, J.; Zhang, G.-F.; Zhang, J.-X.; Na, S.-J. Effects of Power Modulation, Multipass Remelting and Zr Addition Upon Porosity Defects in Laser Seal Welding of End Plug to Thin-Walled Molybdenum Alloy. *J. Manuf. Process.* **2019**, *41*, 197–207. [CrossRef]
12. Xie, M.X.; Li, Y.X.; Shang, X.T.; Wang, X.W.; Pei, J.Y. Microstructure and Mechanical Properties of a Fiber Welded Socket-Joint Made of Powder Metallurgy Molybdenum Alloy. *Metals* **2019**, *9*, 640. [CrossRef]
13. Xie, M.-X.; Li, Y.-X.; Shang, X.-T.; Wang, X.; Pei, J.-Y. Effect of Heat Input on Porosity Defects in a Fiber Laser Welded Socket-Joint Made of Powder Metallurgy Molybdenum Alloy. *Materials* **2019**, *12*, 1433. [CrossRef] [PubMed]
14. Pan, T.-L.; Wang, T.-L.; Pan, M.-L.; Le, H.-L.; Le, C.-T. Electron-beam welding of molybdenum. *J. Tsinghua Univ. (Sci. Technol.)* **1964**, *2*, 15–34.
15. Yang, Q.; Zhu, Q.; Wang, L.; Wang, N. Microstructure and Properties of Mo Plates by Electron Beam Welding. *Hot Work. Technol.* **2014**, *13*, 158–161.
16. Zheng, W.; Liu, J.; Ma, Z. Vacuum electron beam welding trait of 16mm Molybdenum. *Hot Work. Technol.* **2012**, *41*, 153–154.
17. Morito, F. Tensile properties and microstructures of electron beam welded molybdenum and TZM. *J. Less Common Met.* **1989**, *146*, 337–346. [CrossRef]
18. Morito, F. Characteristics of EB-weldable molybdenum and Mo-Re alloys. *JOM* **1993**, *45*, 54–58. [CrossRef]
19. Morito, F. Weldability and fracture of molybdenum-rhenium welds. *Met. Powder Rep.* **1998**, *53*, 46. [CrossRef]
20. Morito, V.N. Slyunyaev Impurity-induced embrittlement of heat-affected zone in welded Mo-based alloys. *Int. J. Refract. Met. Hard Mater.* **1997**, *15*, 325–339.
21. Stütz, M.; Oliveira, D.; Rüttinger, M.; Reheis, N.; Kestler, H.; Enzinger, N. Electron Beam Welding of TZM Sheets. *Mater. Sci. Forum* **2016**, *879*, 1865–1869. [CrossRef]
22. Chen, G.; Liu, J.; Shu, X.; Zhang, B.; Feng, J. Study on microstructure and performance of molybdenum joint welded by electron beam. *Vacuum* **2018**, *154*, 1–5. [CrossRef]
23. Chen, G.; Yin, Q.; Chen, G.; Zhang, B.; Feng, J. Beam deflection effects on the microstructure and defect creation on electron beam welding of molybdenum to Kovar. *J. Mater. Process. Technol.* **2019**, *267*, 280–288. [CrossRef]
24. Wang, H.; Zhang, Y.P.; Zhang, X.Y. TZM molybdenum alloy TIG welding process and microstructure of welded joints. *Heat Treat. Met.* **2012**, *37*, 41–44.
25. Qinglei, J.; Yajiang, L.; Puchkov, U.; Juan, W.; Chunzhi, X. Microstructure characteristics in TIG welded joint of Mo-Cu composite and 18-8 stainless steel. *Int. J. Refract. Met. Hard Mater.* **2010**, *28*, 429–433. [CrossRef]
26. Wang, H. Effects of weld doping on weldability of commercial purity molybdenum. *China Molybdenum Ind.* **1995**, *6*, 20–22.
27. Matsuda, F.; Ushio, M.; Nakata, K.; Edo, Y. Weldability of molybdenum and its alloy sheet (report i) (materials, metallurgy, weldability). *Trans. Jwri* **1979**, *8*, 217–229.
28. Kolarikova, M.; Kolarik, L.; Vondrous, P. Welding of thin molybdenum sheets by EBW and GTAW. In Proceedings of the 23rd Ingernational DAAAM Symposium 2012, Zadar, Croatia, 24–27th October 2012; Volume 23, pp. 1005–1008.
29. Liu, P.; Feng, K.Y.; Zhang, G.M. A novel study on laser lap welding of refractory alloy 50mo–50re of small-scale thin sheet. *Vacuum* **2016**, *136*, 10–13. [CrossRef]
30. An, G.; Sun, J.; Sun, Y.; Cao, W.; Zhu, Q.; Bai, Q.; Zhang, L.-J. Fiber Laser Welding of Fuel Cladding and End Plug Made of La2O3 Dispersion-Strengthened Molybdenum Alloy. *Materials* **2018**, *11*, 1071. [CrossRef]
31. Zhang, L.-J.; Wang, C.-H.; Zhang, Y.-B.; Guo, Q.; Ma, R.-Y.; Zhang, J.-X.; Na, S.-J. The mechanical properties and interface bonding mechanism of Molybdenum/SUS304L by laser beam welding with nickel interlayer. *Mater. Des.* **2019**, *182*, 108002. [CrossRef]

32. Zhang, L.; Lu, G.; Ning, J.; Zhang, L.; Long, J.; Zhang, G. Influence of Beam Offset on Dissimilar Laser Welding of Molybdenum to Titanium. *Materials* **2018**, *11*, 1852. [CrossRef]
33. Zhang, L.-J.; Lu, G.; Ning, J.; Zhu, Q.; Zhang, J.-X.; Na, S.-J. Effects of minor Zr addition on the microstructure and mechanical properties of laser welded dissimilar joint of titanium and molybdenum. *Mater. Sci. Eng. A* **2019**, *742*, 788–797. [CrossRef]
34. Lin, L.; Huang, Y.; Hao, K.; Chen, Y. Cracking in dissimilar laser welding of tantalum to molybdenum. *Opt. Laser Technol.* **2018**, *102*, 54–59.
35. Ning, J.; Hong, K.; Inamke, G.; Shin, Y.; Zhang, L. Analysis of microstructure and mechanical strength of lap joints of TZM alloy welded by a fiber laser. *J. Manuf. Process.* **2019**, *39*, 146–159. [CrossRef]
36. Zhang, L.-J.; Liu, J.-Z.; Bai, Q.-L.; Wang, X.-W.; Sun, Y.-J.; Li, S.-G.; Gong, X. Effect of preheating on the microstructure and properties of fiber laser welded girth joint of thin-walled nanostructured Mo alloy. *Int. J. Refract. Met. Hard Mater.* **2019**, *78*, 219–227. [CrossRef]
37. Gao, X.-L.; Li, L.-K.; Liu, J.; Wang, X.-Q.; Yu, H.-K. Effect of laser offset on microstructure and mechanical properties of laser welding of pure molybdenum to stainless steel. *Int. J. Refract. Met. Hard Mater.* **2020**, *88*, 105186. [CrossRef]
38. Xu, J.; Jiang, X.; Zeng, Q.; Zhai, T.; Leonhardt, T.; Farrell, J.; Umstead, W.; Effgen, M.P. Optimization of resistance spot welding on the assembly of refractory alloy 50Mo–50Re thin sheet. *J. Nucl. Mater.* **2007**, *366*, 417–425. [CrossRef]
39. Ferrenz, E.E.; Amare, A.; Arumainayagam, C.R. An improved method to spot-weld difficult junctions. *Rev. Sci. Instruments* **2001**, *72*, 4474–4476. [CrossRef]
40. Xia, C.; Wu, L.; Xu, X.; Zou, J. Phase constitution and fracture analysis of vacuum brazed joint of 50Mo-50Re refractory alloys. *Vacuum* **2017**, *136*, 97–100. [CrossRef]
41. Song, X.G.; Tian, X.; Zhao, H.Y.; Si, X.Q.; Han, G.H.; Feng, J.C. Interfacial microstructure and joining properties of titanium-zirconium-molybdenum alloy joints brazed using ti-28ni eutectic brazing alloy. *Mater. Sci. Eng. A* **2015**, *653*, 115–121. [CrossRef]
42. Fu, L.; Du, S. On exploring better friction welding joint of TZM Mo-Base powder alloy and H11 Mold Steel. *J. Northwestern Polytech. Univ.* **2001**, *19*, 557–561.
43. Fujii, H.; Sun, Y.; Kato, H. Microstructure and mechanical properties of friction stir welded pure Mo joints. *Scr. Mater.* **2011**, *64*, 657–660. [CrossRef]
44. Reheis, N.; Tabernig, B.; Kestler, H.; Sigl, L.S.; Pretis, D.D.; Enzinger, N. Friction Welding of TZM Components. In Proceedings of the World Congress on Powder Metallurgy and Particulate Materials, Orlando, FL, USA, 18–22 May 2014.
45. Ambroziak, A. Friction welding of molybdenum to molybdenum and to other metals. *Int. J. Refract. Met. Hard Mater.* **2011**, *29*, 462–469. [CrossRef]
46. Stütz, M.; Wagner, J.; Reheis, N.; Kestler, H.; Raiser, E.; Enzinger, N. Rotary Friction Welding of Large Molybdenum Tubes. In Proceedings of the International Conference on Trends in Welding Research, Tokyo, Japan, 11–14 October 2016.
47. Stütz, M.; Pixner, F.; Wagner, J.; Reheis, N.; Raiser, E.; Kestler, H.; Enzinger, N. Rotary friction welding of molybdenum components. *Int. J. Refract. Met. Hard Mater.* **2018**, *73*, 79–84. [CrossRef]
48. Stütz, M.; Buzolin, R.; Pixner, F.; Poletti, C.; Enzinger, N. Microstructure development of molybdenum during rotary friction welding. *Mater. Charact.* **2019**, *151*, 506–518. [CrossRef]

© 2020 by the authors. Licensee MDPI, Basel, Switzerland. This article is an open access article distributed under the terms and conditions of the Creative Commons Attribution (CC BY) license (http://creativecommons.org/licenses/by/4.0/).

MDPI
St. Alban-Anlage 66
4052 Basel
Switzerland
Tel. +41 61 683 77 34
Fax +41 61 302 89 18
www.mdpi.com

Metals Editorial Office
E-mail: metals@mdpi.com
www.mdpi.com/journal/metals

www.ingramcontent.com/pod-product-compliance
Lightning Source LLC
LaVergne TN
LVHW070718100526
838202LV00013B/1122